INTO THE LIGHT

Tomorrow's
Medicine Today

INTO THE LIGHT

Tomorrow's
Medicine Today

William Campbell Douglass, M.D.

Second Opinion Publishing Inc.
Dunwoody, Georgia 30338

Copyright © 1993

by

William Campbell Douglass, M.D.

ISBN 0-9626646-5-0

Library of Congress Catalog Card Number
91-067330

Cover illustration by Bill Buford

For information regarding this book, call or write:

Second Opinion Publishing Inc.
Suite 100, 1350 Center Drive
Dunwoody, Georgia 30338
(800) 728-2288

Explanatory Note

The medical process of photo-oxidation, about which much of this book is written, is known by many names. Because of the necessity to quote from many original manuscripts, all of the following terms will have to be used interchangeably to indicate the photo-oxidative process (followed parenthetically by the names of those who used them predominantly):

Hemo-irradiation (Eidam)
Photoluminescence (Douglass)
Photopheresis (Edelson)
Photochemotherapy (Rebbeck)
Photobiological therapy (Douglass)
Photo-oxidation (Douglass)
Ultraviolet blood irradiation, or UBI (Olney, Knott, Miley, et al)
Photon pump (Gallagher)
Photodynamic therapy (Matthews)

Dedication

To Emmett K. Knott, pioneer in photo-therapeutics.

To all the doctors of Equatorial Africa – who work so hard, with so little – always hoping, never despairing.

And to my Russian colleagues in physics and medicine who work for the love of science and the satisfaction of having helped mankind.

Contents

Section III: Other Faces of Light Therapy

The Tao of Light

The modern theory of relativity and many magical discoveries concerning light have brought about a renaissance in philosophical thinking. It is now realized that there is no such thing as "exact science," and everything is in a sphere of probability. As Dr. Johanna Budwig has remarked, "It is not the worst scientist who emphasizes the limits of all scientific findings. It is definitely no coincidence that, especially among the best and most intelligent physicists, the Christian conviction is now strongly emphasized."

The discovery of quantum physics should have led to quantum biology, but this is little understood or practiced in modern medicine. Instead, "molecular biology," in which the "building stones" of life and life's processes are studied, attempts to solve life's problems and mysteries through this highly complex and yet simplistic approach to life.

So when you get down to the real nitty gritty of reality — life and the Creator — you're talking about light, which is believed to be the smallest "particle." But this "particle" has no mass, and therefore exists only in the mind through the perception of the brain.

If you combine two of these massless "non-existent" light particles, then you have an electron which has mass and weight; you can "pick it up," so to speak. You have passed from the unreal world of quantum mechanics to the material world as we view it. How does light get transformed into physical substance? How do two nothings become a something? That's God's little secret, and I doubt He will ever reveal it. There are limits to what man can understand — at least I think so.

Certainly no one worries about the relevance of quantum mechanics when blowing his nose, repairing a car or imploring God to forgive previous sins. But this fussy, unrealistic world of quantum mechanics reminds us of how little we know and how, through the omnipresence of light, there is also the omnipresence of God. If you influence a light particle here, and its twin a universe away can be affected by that influence, where does that leave us? (Don't ask me.)

On a more prosaic note – although scientists don't have the slightest idea what's going on when they get inside the atom, they can do remarkable things with new light technology.

John Tyndall, a brilliant British physicist, started the whole thing back in the 1870s when he demonstrated that light could bend. Ten years later, Alexander Graham Bell proved that a light beam could carry voice signals. It took 80 years for the significance of this to be fully appreciated, and, in the 1960s, fiber-optic science was born. As we close the 20th century, we can, through using very tiny spider wed-like strands of glass, send out to the world 1.8 billion bits of information a second. Sophisticated lasers, with a light source no bigger than a grain of salt, can blink on and off about one billion times a second. We can now feed several signals into one of these glass strands at the same time and can even carry signals in both directions at the same time. Your telephone company, through a single glass strand, can transmit 16 thousand phone conversations at once. This compares with 24 conversations capable by the old fashioned copper wire technology. The entire text of the Encyclopedia Britannica and the Bible could travel around the earth through a fiber cable in less than two seconds.

This technological revolution in the use of light, although we don't understand it, is now applicable to the practice of medicine and will revolutionize both diagnosis and treatment.

If you don't believe in a Creator, perhaps you will after reading what scientists are saying these days about light.

Everything in the sciences seems eventually to come down to light. The world appears neat and orderly from a distance. Energy seems to be separated into waves or particles, but close scrutiny of light, the ultimately small particle, reveals indescribable objects that have characteristics of both.

As scientists studied atoms, a nucleus of solid matter surrounded by circling electrons, they realized that matter is very strange, full of uncertainty, and, like the Hindu Veil of Maya, the closer you get into the world of the atom, the more illusory reality becomes – it actually exists only in the eyes of the observer. Einstein was perhaps the first to recognize this uncertainty, and it bothered him greatly. He said, "God does not play dice with the universe" (which still leaves us in the dark).

The study of the subatomic world, i.e., the world inside the atom, is called quantum mechanics. Quantum mechanics defies common sense and illustrates phenomena that seem impossible in the real world. The mind cannot conceive of the postulate that goes with quantum mechanics: a change in an object, such as a light particle, can instantly produce a change in a related light particle – even on the other side of the universe.

The only way that man can grasp this strange and unreal world is through the study of photons, the basic unit of light. John Clauser, who works for Bell Laboratories, studied photons and proved that a change in one photon did alter the polarization of another. Photons travel in pairs, called biotrons. These two related, paired photons, Clauser says, cannot be considered separate objects, but rather remain connected in some very mysterious way. This finding, says physicist Henry Stapp of Lawrence Berkley Laboratories, "imposes new limits on what can be established about the nature of matter by proving that experiments can be influenced by events elsewhere in the universe."[1]

Conventional medicine has a mechanistic and compartmentalized view of the body. Investigational techniques used today illustrate this approach: X-ray, body scanning techniques, nuclear magnetic resonance, etc. We can see more and more structure in more and more detail. But these *structural* findings tell us nothing about *functional* change.

Modern medicine puts the cart before the horse. Usually, irreversible structural change follows functional change that has been in process for months or years. When the diagnosis of modern medicine is "nothing abnormal found," while the patient has obvious serious complaints, this means simply that modern medicine has again failed to make an early diagnosis and must wait, treat the patient's symptoms, and attempt a diagnosis later, after structural change has taken place and the disease is irreversible (coronary heart disease, cancer, gall stones, arthritis deformans, atherosclerosis, etc., etc.).

---

1 *Time* magazine, April 16, 1990.

The old Newtonian view of physics, and the world in general, the mechanistic building block approach to science, is no longer tenable. Quantum science is a unified view of the world that recognizes the interrelatedness of everything, including biological processes. Until medicine catches up to physics, and at present there is very little awareness in medicine and biology of the revolution that has taken place in physics, there will be no change in medicine and little practical progress toward truly early diagnosis at the photo-electronic level and toward holistic therapy.

Assuredly, more and more spectacular scientific discoveries will be made, but from a practical point of view, most of this scientific information won't be worth knowing.

Biocybernetic medicine is quantum physics applied to biological organisms. It recognizes the interrelatedness and intercommunication between every cell in the body. It recognizes that in order to diagnose and treat early, this communication system must be tapped into. This network goes far beyond what we generally think of as the nervous system and includes the ancient acupuncture points and meridians and energy fields that we can measure but do not completely comprehend.

This biocybernetic or bioenergetic medicine takes many forms, the oldest being Chinese acupuncture. A more sophisticated form of biological energy regulation is the practice of homeopathy. But the greatest advance in medicine in the twentieth century has been the wedding of oxygen enhancement (bio-oxidation) and light therapy (photoluminescence). This combined therapy, which we call photo-oxidation, has truly brought medicine into the real world of quantum mechanics.

It is interesting to note that conventional medicine is the only system of medicine that has no concept of biological energy — that's why today's medicine is intellectually impotent, treats only symptoms, and must resort to surgery, drugs, and X-radiation as a result of its inability to make an early diagnosis[2] and then use biocybernetic methods of treatment. Conventional medicine will not use nontoxic, bioenergetic methods, such as

---

2 By "early" we don't mean a tumor that can be seen on an X-ray, an abnormal lab test or a positive biopsy — these are *late* manifestations of disease.

photo-oxidation, because conventional medicine has no concept of altering biological energy flow.

Ilya Prigogine, a brilliant Belgian chemist, won the Nobel Prize in 1977 for his theory of dissipative structures. Prigogine disproved, through some complex mathematics, the second law of thermodynamics, at least in relation to biological systems.

The second law of thermodynamics states that the universe, and everything in it, is running down or dying (entropy). But Prigogine proved that biological systems, including man, can grow rather than dissipate through disease processes followed by recovery. In other words – an energy disruption (disease), when corrected by photo-bioelectric methods (acupuncture, homeopathy, neural therapy, photo-oxidation, etc.), enables the organism to reach a higher state of health.

Obviously this is only possible through holistic (photo-biocybernetic) medicine, because only through these sophisticated methods can disease (photo-bioelectric disturbances) be detected early and corrected before permanent damage (entropy) has taken place.

Prigogine's principle confirms the value of periodic visits to a biocybernetic physician. The body is in a constant state of disease and recovery. Cancer, viral and bacterial invasions, and other disruptions, are an everyday occurrence. The healthy body corrects these problems and builds to a higher order of health. As one gets older these conditions need early correction to avoid clinical disease. Only through photo-biocybernetic medicine can this be done successfully. This is preventive medicine in the truest sense of the word.

The future of photobiology and photochemotherapy is indeed bright. The present would be bright, too, if the American people only knew what could be done with photochemotherapy in the form of photo-oxidation. That, of course, is the purpose of this book.

Section 1

Light Therapy:
A Long, Unqualified
Success Story

Chapter 1

Light: Prerequisite for Health

Light and health have always gone together. Though only recently have we begun to understand some of the reasons why light is so beneficial, men have always instinctively sought it. Ancient men may not have known that exposure to light helped their skin utilize vitamin D, or that light is an essential nutrient, just like vitamin C. But they did know that there was no therapy more helpful than walking in a field of clover on a beautiful spring day, especially after being indoors through a long winter. Both the mind and the body were lifted and enriched and healed.

That being the case, perhaps it should come as no surprise to us that light has even greater implications for health when coupled with modern medical science. If light has inherent qualities that promote health, focusing those qualities where they are most needed through variable-reducing scientific techniques should provide for even greater healing.

Following such logic, medical scientists pioneered amazingly effective techniques of light therapy early in this century.

Using the ultra-violet portion of the light spectrum, these pioneers were able to cure many previously hopeless diseases. In fact, an incredible variety of afflictions were cured using their techniques. People near death, with formerly untreatable conditions, were remarkably restored to complete health. The future of light therapy seemed assured. At about the same time, however, antibiotic treatment made its debut. Since antibiotics also treated numerous diseases successfully, and since they had the additional advantage (to the medical community) of requiring the production of prescription drugs, they soon became the treatment of choice.

Today, photoluminescence is all but unknown. As this book will attempt to demonstrate, however, it is just as exciting and revolutionary today as it was half a century ago. Proper use of light therapy holds health implications beyond the imagination. It even offers very real hope for the treatment of AIDS. Actual in-the-field clinical treatment of AIDS patients in Africa has yielded enormously promising results in a high number of patients. Other modern maladies that resist known therapy have also responded well to photoluminescence. Just as important, the side effects are minimal. With the proliferation of new diseases in our modern society, it is truly time to fully exploit this most effective medical technique.

When any new medical treatment comes on the scene, and is said to cure practically everything, it is looked upon by conventional medicine with extreme suspicion. This is certainly understandable. Antibiotics were *supposed* to stamp out infectious disease; cortisone was *supposed* to stamp out allergies; immunizations were *supposed* to stamp out infectious disease entirely – *none* of this has happened. There are more people in hospitals with infectious diseases than ever before in modern times. Allergies are rampant. None of the panaceas have worked.

So when a new mode of treatment comes on the scene, it is greeted with an understandable suspicion: "If it is supposed to cure everything, then it probably cures nothing." This is why photoluminescence is such an amazing paradox. It is a tested and proven therapy which has accomplished incredibly miraculous cures, with virtually no side effects, and yet, until recently, has all but disappeared from the medical scene.

The operational strength of photoluminescence is its administrative weakness, however. As a curative agent, it requires little sophisticated equipment, no complicated drugs, and it cures by stimulating the body's own immune response (the key to light therapy's versatility). Such a medical breakthrough, while a windfall for the ailing public, is a death knell for some members of the medical establishment – money in the consumer's pocket is money that never makes it to the medical-industrial complex.

My intent is not to dwell on intrigue, however. The foregoing is mentioned only to provide some explanation of why such a curative technique has not made banner headlines in medical journals or public newspapers. My chief purpose is rather to expound on the nature and efficacy of this near-miraculous therapy, with the hope that it will take its rightful place in medical practice.

What is Light Therapy?

The most appropriate name for light therapy is probably "photoluminescence." So extensive has the research been in this area, and so varied the researchers, that it has been called by several names. Among these are hemo-irradiation, photopheresis, photochemotherapy, photobiological therapy, photo-oxidation, ultraviolet blood irradiation, photon pump, and photodynamic therapy. These names are instructive in themselves, giving some idea of the scope and diversity of photoluminescent research in the past.

The word *photoluminescence* sounds mysterious and hi-tech, but it is an extremely simple, painless, and safe means of treating a patient. "Photo" refers to light, and "luminescence" refers to emission of light. Although this book deals primarily with the ultraviolet portion of the energy spectrum, other energies, from x-ray through infrared, and perhaps beyond, can be applied to the treatment of disease. Starting with UV light, we are opening a veritable cornucopia of therapeutic possibilities using the electromagnetic forces that surround us.

Ultraviolet light has been used in disinfection for many years and is, in fact, still used for that purpose. Any contaminated object, whether it be surgical instruments, bedding, room air, the human skin, or body fluids such as blood, can be cleansed rapidly of viruses and bacteria.

This killing of infectious organisms is a useful quality of ultraviolet light, but it is not as important as another capability of this remarkable part of the energy spectrum: *the stimulation of the immune system and various enzyme systems.*

When a small quantity of blood is treated through photoluminescence, an astounding thing happens. Through some mechanism that is not completely understood, the body's defenses are organized rapidly to destroy all invading organisms, whether viral, fungal, or bacterial. The immune system comes to life and rapidly brings the body back to a state of balance.

To illustrate this phenomenon, allow me to describe two very typical cases:

Don Pool drove a hundred miles into the mountains of Georgia to bring his sister, Patrice, to see me for treatment. She had a typical case of "the crud" with sneezing, coughing, aching, sore throat, weakness, and extreme fatigue – typical "flu."

In a simple ten-minute procedure, I treated her with photoluminescence. Don had driven his sister to my clinic and so was able to observe her on the way back to their home in Atlanta. He said that about an hour after leaving my office, halfway back to Atlanta, she developed a severe aching all over and a definite fever. She felt worse than she had before the treatment. But by the time they reached home, an hour later, she was completely well. Her runny nose, her cough, her aches and pains, her sneezing and her malaise were completely gone. They simply could not believe it.

Don began to experience the same symptoms a few days later. Because he had not taken my advice to be treated when he was at my office with his sister, he drove back up to the mountains to see me for treatment. His symptoms were not quite as bad as Patrice's, but he had obviously caught the same flu virus. We treated him in exactly the same way; he drove home and experienced the same type of crisis as his sister. He called me the next day to say that he, too, had completely recovered.

There is much in current scientific literature about a new cure for the common cold, using a molecule to which the cold virus attaches. This molecule, called ICAM-I receptor, is thought to be effective in the treatment of colds if sprayed into the nose.

The molecule, made synthetically and sprayed onto the mucous membranes of the nose, confuses the invading cold virus. This would, presumably, lure the cold virus away from the real cell receptors of the body. Once bound to the synthetic molecule, the virus particles would be neutralized and thus be unable to cause a cold. Researchers admit that this probably would not cure an active cold, but might be used as a preventive measure. Thus far, it has only worked in the test tube. Relief from this technique is still years away.

Photoluminescence, however, is here now, cures colds in the same manner as it did the flu in the above cases, and costs very little. Why wait for the expensive, complicated, and unproven technology of ICAM-I?

Although discovered 60 years ago, photoluminescence has not been used as a treatment method for infections since the 1940s. Since antibiotics were to be the panacea for all infectious diseases, widespread use of other techniques has never been realized. Although antibiotics have been a great benefit to mankind and have saved millions of lives, what antibiotics have accomplished has been at considerable cost.

Part of that cost has been severe yeast infections, depression of the immune system, replacement of targeted infectious organisms with others not previously thought to be capable of causing disease, a dramatic increase in various degenerative diseases, and a general increasing resistance to the effects of antibiotics. Consequently, we now have penicillin-resistant syphilis, drug-resistant gonorrhea and incurable tuberculosis.

Clearly, a new breakthrough is needed to counteract the incredible, awesome, and terrifying increase in infectious diseases that we are now facing. Not only do we have treatment-resistant syphilis, gonorrhea, and tuberculosis, we are also facing a new venereal disease called chlamydia; the new Lyme disease; the return of dengue fever; a new and deadly mycoplasma (organisms larger than viruses, but smaller than bacteria) infection; a veritable epidemic of candidiasis (a yeast); and of course, Acquired Immune Deficiency Syndrome – AIDS. We are now in what could be called the New Age of Pestilence.

Photoluminescence is the medicine of the future – and it is here now. It's *been* here since the 1930s, but doctors seem to be able to concentrate on only one line of medicine at a time. So photoluminescence was discarded along with many other nontoxic, effective treatments that did not fit into the traditional philosophy of therapy.

Originally, the method of exposing blood to light, or "hemoirradiation," was developed to utilize the bacteria-killing properties of ultraviolet rays by directly irradiating the bloodstream, and thereby combat bloodstream infection. This concept was successfully developed by several researchers. Clinical observations on patients treated by this method, however, indicated that several *additional, decidedly beneficial* biological and biochemical reactions were taking place within the patient. One researcher, Dr. Virgil K. Hancock, listed several of these surprise effects as follows:[1]

---

1 *American Journal of Surgery*, December 1942.

1. An inactivation of toxins and viruses.[2]
2. Destruction and inhibition of growth of bacteria.[3]
3. Increase in the oxygen-combining power of the blood.[4]
4. Activation of steroids.[5]
5. Increased cell permeability.[6]
6. Absorption of ultraviolet rays by the blood and emanation of secondary irradiations.[7]

Dr. E. W. Rebbeck noted that in the over 3,000 blood irradiations that he had given, there was often a dramatic increase in the number of red cells in the blood. This increase was often as high as 1.2 million cells overnight. He noted that the white cell count often would tend to balance – high counts would decrease and low counts increase to a normal level.

Effects even beyond those listed by Dr. Hancock have been observed. These include:

1. Activation of cortisone-like molecules, called sterols, into vitamin D.[8]
2. Bacteria in the body are killed directly by ultraviolet light and indirectly by increasing local and systemic resistance. This effect is not entirely understood.[9]
3. Ultraviolet radiation tends to restore normal chemical balances in the body.
4. In suitable doses it tends to correct cellular imbalance in the blood.
5. Fat elements in the blood, which are altered in character by disease, are restored to normal size and movement by ultraviolet light.

---

2 Macht, "Contributions to Photopharmacology or the Applications of Plant Physiology to Medical Problems," *Science*, 71: 303-304, March 21, 1930.

3 Zinsser, et al, *Textbook of Bacteriology*, New York: Appleton Century Company, Inc., 1939.

4 Miley, "The Ultraviolet Irradiation of Auto-Transfused Human Blood," Studies of the absorption value. *American Journal of Medical Science*, Vol. 14, 1938-1939.

5 Duggar, *Biological Effects of Radiation*, New York: McGraw-Hill Book Company, p. 323-335.

6 Ibid.

7 Luckiesh, *Light and Health*, Williams and Wilkins Company, 1926, P. 66-91.

8 Duggar, p. 323-335.

9 Ellis and Wells, *The Chemical Action of Ultraviolet Rays*, New York: The Chemical Catalog Company, 1925.

6. Ultraviolet radiation has cumulative effects; each treatment builds on, and enhances, the effects of previous treatments.
7. Individuals vary greatly in their sensitiveness to ultraviolet irradiation regarding systemic and skin effects.
8. Sensitivities may be modified by certain drugs, such as sulfanilamide; foods, such as buckwheat; and substances, such as hematoporphyrin (a building block of hemoglobin, which is the oxygen-carrying molecule in the blood).
9. Overdosage with ultraviolet produces depression, lessened resistance to bacterial infections and reduced bacteria-killing potency of the blood with a fall in hemoglobin. The level of exposure required for an overdose is never approached in proper clinical practice.
10. The action of ultraviolet may be immediate, somewhat delayed, markedly delayed, or protracted.

This list of effects does not completely explain the complicated physiological action and effects of ultraviolet light therapy. It may be years before this mystery is unraveled but the therapeutic miracle it represents can no longer be ignored.

Ultraviolet rays have also shown, repeatedly, the ability to inactivate virus toxins. The work on snake venom in Japan by Dr. Noguchi, who inactivated both cobra and rattlesnake venom, is one example of this remarkable capability.[10] Tetanus, botulism, and rabies toxins have also been inactivated by ultraviolet blood irradiation. (See Chapter 4).

This important phenomenon is also evident in photoluminescent therapy, due to this general action of ultraviolet light. In cases of serious infections, a marked reduction of toxic symptoms is noticed within 12 to 48 hours after treatment. Malaise, severe headaches, nausea, mental confusion, chills, fever, and other toxic symptoms are usually quickly relieved.

It is interesting that less ultraviolet exposure is required to kill bacteria in the human body than is necessary in the laboratory. When a small part of the infected bloodstream is exposed to ultraviolet light for *less time than is required to kill bacteria in*

---

10 Noguchi, *Journal of Experimental Medicine*, 252-267, 1906.

the laboratory, the bacteria in the body are usually *completely destroyed.* In fact, many organisms are destroyed by an amount of irradiation that merely *stimulates* normal body cells.

At least part of this bactericidal phenomenon may be explained by two aspects of the chemical structure of bacteria. First, the body cells are immense compared to bacterial cells. The impact of ultraviolet light exposure can therefore be expected to affect them each differently. Second, certain bacteria contain significant amounts of two photo-sensitive amino acids – phenylalanine and tyrosine – that are almost absent in body cells. Because of their photo-activity, these amino acids cause the bacteria to absorb greater amounts of ultraviolet energy. This absorption of energy begins a coagulation process which destroys the bacteria. Since body cells are not as photosensitive, they are not affected. In fact, since bacteria contain at least five times as much of these amino acids as human cells, they absorb five times as much photonic energy from the UV-treated blood. There is also some reason to believe that by destroying the bacteria in the treated sample of blood, an "autogenous [self-generated] vaccine" is produced. When this vaccine is coupled with radiation given off by the treated blood – "induced secondary radiation" – the bacteria in the patient's bloodstream are rapidly destroyed.

Induced secondary radiation is a known physical phenomenon that is easily observed in human blood that has been exposed to ultraviolet light. Hemoglobin absorbs radiated light energy over a wide range of wave lengths. By placing irradiated blood in a thin-walled quartz vessel and setting it on an ultraviolet-sensitive film, the film will be quickly fogged. This fogging is evidence of radiation being *given off* by the blood luminescence. Such secondary radiation is probably the primary mechanism that causes the remarkable effects observed in photoluminescent therapy.

Dr. George Miley was a clinical physician who practiced photoluminescence extensively in the 1930s.[11] His work clearly demonstrated another important effect of light therapy – an

---

increase of oxygen absorption by the blood following ultraviolet exposure. This effect has a dramatic impact on the patient, particularly in the presence of cyanosis (blueness) from lack of oxygen. When a patient is cyanotic from the administration of a drug such as sulfanilamide or from some other condition, photoluminescent treatment usually causes a return to healthy color within a few minutes.

Dr. Miley also reported on 151 consecutive unselected cases of acute infection treated by photoluminescence.[12] The results were phenomenal. In those cases that were treated early, 100 percent of the patients recovered fully. In moderately advanced cases, 98 percent recovered; and even patients who were moribund (near death) experienced a 42 percent recovery rate. (See Chapter 6.)

Dr. E. W. Rebbeck also practiced photoluminescence earlier this century. He reported on 13 patients with septicemia (blood poisoning) following childbirth. All 13 recovered following photoluminescent treatment. Six of these patients had been treated previously with sulfa drugs, to no avail. These remarkable results prompted Dr. Rebbeck to use hemo-irradiation for all patients suffering from blood poisoning following childbirth. He also did a study on 16 patients suffering from septicemia following abortion. All 16 recovered fully. Seven of these patients were in a state of "advanced morbidity" when treated, and had not been expected to live. All of the patients had been treated with sulfa drugs previously, without success. (See Chapter 6.)

Another researcher, Dr. Virgil K. Hancock, reported a case of bloodstream infection in which the patient's temperature had risen to 108.4° F. The patient also had a bone marrow infection and bronchial pneumonia. In those days (about 1938), this case would have been considered essentially hopeless. The patient made a complete recovery.

The visible portion of the light spectrum can also be used for healing, and a great deal of research has been done on this subject, as well. (In the last section of this book, I'll review some of this research and detail some of the therapeutic methods of using light other than ultraviolet.) Dr. Morton Walker has written

---

12 Miley, G., *American Journal of Surgery*, 73: p. 486-493, 1947.

a fascinating book called *The Power of Color* in which he enumerates the various therapeutic potentials of the different parts of the rainbow. He states that "whether or not the ... physician accepts or is even cognizant of this reality of color therapy, color is basic to any system of healing. Color is nature's own curative measure...."

Walker's research on color therapy revealed that the color red can be used to treat anemia and constipation, but is not recommended for hypertension and mental illness. The color orange can be used in the treatment of gout. Green is effective in the treatment of cancer; blue is helpful for itching and insomnia, though ineffective for treating colds and gout; and violet is good for mental illness and cancer.

The Walker book is fascinating reading and I recommend it highly.[13] However, I do not recommend color therapy to the exclusion of full-spectrum light. (See Chapter 16.) You really need to consult a doctor who knows a little about photobiology to get the full benefits of light therapy. *Good luck!*

What you will likely find instead of a light therapist are plenty of the physicians Dr. Albert Wahl described. Concerning new or unorthodox therapies, he said:

> The usual reaction of physicians who are willing to indulge in the dubious intellectual luxury of condemnation without investigation – of which I myself was formerly guilty – is, that if the patients get better, they either didn't have the disease, or they are neurotics; if they do not improve, the medicine is no good; if they die, the medicine killed them. May I say hypothetically, that if this is merely a cure for neurotics, it should be welcomed....[14]

---

13 Avery Publishing Group, 120 Old Broadway, Garden City Park, New York 11040. Tel.:1-800-548-5757.
14 Albert L. Wahl, M.D., public remarks, Summer 1947.

Chapter 2

History of
Ultraviolet Therapy

That ultraviolet therapy is so useful in medical applications should really be no surprise. It has long been recognized and used by the medical profession in the treatment of various disorders. For an even longer period of time it has been used as a method of sterilization in medicine and in certain commercial applications. In the "old days" some washing machines had a built-in ultraviolet light. The light was a promotion technique because it was common knowledge that UV had an anti-bacterial effect. It was also a common practice to irradiate public toilet seats with ultraviolet light.

In fact, the existing body of knowledge on ultraviolet therapy is voluminous, thorough, distinguished, and time-tested. Anyone interested in researching the subject can have access to an incredible amount of scientific knowledge and research from the work of a hundred-year period.

Niels Ryberg Finsen was the father of scientific ultraviolet radiation therapy. Through his work in the late 19th century he brought the healing action of this form of radiant energy to the attention of the medical world. He began his work on the ultraviolet treatment of skin and mucus membranes as far back as the 1880s. Finsen and his successors treated more than 2,000 patients with various skin conditions. According to these researchers, the success rate was 98 percent. Their cases primarily entailed the treatment of lupus vulgaris, a tuberculosis-like disease of the skin and mucus membranes. The credibility of his work extended beyond mere numbers and percentages, however. He was awarded a Nobel prize in 1903 for his photo-chemotherapeutic work. Then, in 1904, Queen Alexandra made a royal visit to a clinic at the London Hospital, Whitechapel, where the new method was being used. Her presence there gave an even further boost to the credibility of the therapy and its inventor. As a result, many researchers since Finsen have explored the physiological effects of ultraviolet spectral energy, as well as its curative effects on many diseases.

Because of such labors, there exists an extensive body of literature dealing with all aspects of the subject. Ultraviolet therapy has proven to be an immensely valuable gift to mankind for the prevention and alleviation of many diseases and for the maintenance of sound health. But a gift must be received to be of any use to the intended beneficiary.

Unfortunately, such reception – on any widespread basis – remains in the future. There exists in the medical profession today an almost total ignorance of ultraviolet therapy. When the subject is mentioned, it almost always draws an antagonistic response. The ignorance is certainly a root cause of the antagonism, but the problem of ignorance extends beyond mere obliviousness of photoluminescence.

Few physicians are well grounded in physics, biophysics, and technical therapeutic procedures involving ultraviolet radiation. They are therefore bereft of the tools needed to understand the basics of the process. While I believe this condition will be forced to change because of the AIDS pandemic, the solution entails first learning basics, and then exploiting the massive body of information on this procedure. As doctors and researchers look in desperation for innovative new weapons to defend us in this new age of pestilence, I have great hope that such educational and attitudinal weaknesses will be overcome.

For the present, however, the neglect of this procedure is certainly one of the century's great tragedies. Untold deaths and much suffering could have been averted if ultraviolet therapy had not been largely abandoned after the discovery of penicillin.

A dramatic example is that of Jim Henson, the creator of the Muppets, who died in 1990. Henson was healthy and had an uneventful medical history, yet he contracted pneumonia and died two days after being admitted to the hospital. Being in the prime of life, Henson did not die of the streptococcus pneumonia for which he was admitted. He died of streptococcal *toxemia* – the virulent toxin that was produced by the strep organism. The great tragedy is that photoluminescent therapy will immediately neutralize such toxins, but was never used.

One of the many diseases that is susceptible to ultraviolet therapy is a skin infection known as erysipelas, which is also caused by a streptococcus. This infection was treated successfully in the 1920s. Dr. Walter H. Ude, of Minneapolis General Hospital, reported a series of 100 cases in which he claimed an almost 100 percent rate of cure with ultraviolet skin irradiation.[1] This method of treating erysipelas was abandoned with the advent of the antibiotic revolution.

Mumps were also proven to be cured easily by ultraviolet treatment – photoluminescence in this case. Though mumps usually causes no permanent damage, it can elicit a great deal of pain and discomfort and, occasionally, sterility in males.

Many researchers agreed that photoluminescence was the perfect treatment for mumps and would probably obviate any serious infection or complications. They found that after only one treatment, the patient's temperature would drop to normal within 24 to 48 hours. Whatever toxic symptoms were present would then be completely cleared in a maximum of 36 hours. The swollen glands under the neck, which are typical of mumps, would disappear within four to five days. And most importantly, "orchitis," swollen testicles which often lead to permanent sterility, could be prevented by early phototherapy.

By the mid-1930s, ultraviolet light was being employed as a systemic treatment through *external* radiation. The application of ultraviolet light to specific disease problems began in the 1940s. Research centers of the time, unlike those of the present day, were very generous in sharing what information they had with private doctors who wanted to apply ultraviolet treatment in a clinical setting. Progress in this new field of phototherapy was therefore made very rapidly.

Moving from *external* to *blood* irradiation with ultraviolet light represented a quantum leap in realizing light's healing potential. When bacteria, fungi, viruses, or protozoa have entered the bloodstream, or have been lodged under even the thinnest layer of tissue, they are protected against the bactericidal effect of outside ultraviolet radiation. The invader can then be destroyed only indirectly by the stimulation of blood cells by ultraviolet

---

1 *Journal of Surgery*, Vol. 61, No. 1, 1943.

light. By exposing a small amount of blood to ultraviolet light over time, researchers found that they could destroy infectious bacteria, viruses, and fungi, no matter how deeply hidden in the body, by stimulating blood cells to inactivate the invaders and their toxins.

One of these pioneers of photoluminescence, Mr. Emmett K. Knott, irradiated the blood of his first human subject in 1928. The patient had a case of sepsis (bloodstream infection) following an abortion. She had been declared beyond help by the attending physicians, but responded dramatically to the irradiation. She recovered rapidly and later bore a normal child.

Working with Dr. Virgil K. Hancock, Mr. Knott presented several historic cases before a meeting of the King County Medical Society in Seattle, Washington on February 5, 1944. These cases made the point, unmistakably, that photoluminescence was a serious, effective method of treatment. In cases of life-threatening infection, proper administration of photoluminescence brought about total recovery when existing methods of treatment were inadequate. These cases are quoted at length in Appendix F and give some idea of the revolutionary, invaluable nature of photoluminescence.

Since these cases occurred before the use of antibiotics (other than sulfa), it is almost certain the patients would have died without ultraviolet, photoluminescent therapy.

Hancock and Knott presented the following explanations of why they found ultraviolet irradiation of the blood to work:

1. Coagulation of bacteria caused by the creation of an "autogenous [self-generating] vaccine."
2. Increased germicidal properties of the blood and an increased number of antibodies (today we would call this effect "stimulation of the immune system").
3. Secondary radiation by treated blood cells (luminescence).
4. Increase of vitamin D content and cholesterol in the blood plasma.
5. Increased oxygen absorption by the blood.

The first article on the therapeutic efficacy of phototherapy was published in June of 1934. Written by Hancock and Knott, it described many of their own cases. Though the article revealed the incredible power of ultraviolet light in the cure of various infections, little attention was paid to it. As E. W. Rebbeck commented in 1941 concerning Hancock and Knott, they "had, in the irradiation of the blood with ultraviolet spectral energy, a therapy of more pronounced value than any other method known to date."[2]

The work of Hancock and Knott took ultraviolet therapy a giant step forward. They had demonstrated that ultraviolet light could be used effectively in the treatment of bloodstream infections. "The toxins in the bloodstream ... exposed to ultraviolet irradiation [are] inactivated. Clinical experiments have further determined [that] the beneficial energy [is] stored up in the rayed blood temporarily, and if such blood can be returned to the bloodstream immediately after it has been irradiated, it will throw off secondary irradiation which will stimulate and energize the patient." By June of 1942, *6,520 patients* had been treated with ultraviolet therapy. Not only had the treatment worked nearly every time, *it had done so in the complete absence of any harmful effects.*

Their work was done before the immune system was clearly understood, so Hancock and Knott had made a monumental breakthrough in therapy without really understanding the processes involved. In spite of our advances in medical knowledge, however, the work of these early pioneers was so far ahead of its time that we still, after 50 years, do not completely understand how photo-irradiation of the blood accomplishes what it does.[3]

The process employed by these researchers was very simple. They removed a small amount of blood, added an anti-coagulant to keep it from clotting, and then irradiated it with ultraviolet light through a window in a closed, air-tight circuit. The blood was immediately returned to the circulatory system of the patient at the same place in the vein from which it had been drawn.

The amount of blood used was calculated on the basis of the patient's vitality, his blood count, the type of infectious agent and the severity of the infection. They would repeat the procedure as often as necessary, but not more than twice a day. Treatment would continue until a sufficient amount of the toxin had been neutralized, and the invading bacteria killed. This allowed the

2 Rebbeck, *Hahnemannian Monthly*, April 1941.
3 *Journal of Surgery*, Vol. 61, No. 1, 1943.

body's normal immune system to take over and finish destroying the foreign organism.

The maximum volume of blood withdrawn by Knott was 1½ milliliters per pound of body weight; an amount which rarely exceeded 300 milliliters. He then exposed the blood to light for an average time of 10 seconds. This procedure contrasts starkly with the method currently in use at Yale University. There, the blood is exposed to light for several hours through the *complete* exchange of the patient's blood supply. Given the phenomenal results Knott and Hancock achieved with much smaller blood volumes, the "Yale Method" appears to be unnecessarily complicated and therefore unnecessarily expensive.

In one of the earlier experiments on photoluminescence, human blood was infected with Staphylococcus aureus bacteria (bacteria that cause "staph infections") and then irradiated with various exposure times of ultraviolet light. Timed exposures of 0, 5, 10, 15, 20, 30, 35, 45, 60, and 70 seconds were used. Cultures were then taken of the samples to determine at which level the ultraviolet had effectively killed all the bacteria in the sample. Only the first three samples showed bacterial growth, so the exposures of over ten seconds killed all the bacteria. Further, there was no apparent effect on the blood cells with exposures up to 70 seconds. After 70 seconds, however, the blood began to show signs of cellular distortion.

This distortion, or "injury," to the cells does not in any way endanger the patient. In fact, one modern researcher theorizes that injured cells are the key to the immune response.[4] But Knott and Hancock used a very small volume of blood in their patients, seldom more than 300 cc, so cell-injury side effects did not occur as with some of the modern techniques.

Knott found in his early work with photoluminescent therapy on dogs that there was a limit to the amount of blood that could be safely irradiated. When too great a volume was exposed, shock and death ensued. Since such a small volume of treated blood is required in order to be effective, this volume restriction is never a factor in clinical practice. However, this finding serves as a warning for such procedures as those

---

4 Edelson, R. L., *Scientific American*, August 1988.

conducted by Yale University where the entire blood volume of the patient is irradiated.

The Forgotten Scientists of Ultraviolet Therapy

In the few instances where modern researchers have paid attention to photoluminescence, they have treated it as a new technology. In doing so, they have virtually ignored the pioneering work of such notable researchers as Knott, Miley, Rebbeck, Barrett, and others. Dr. Richard L. Edelson of the Department of Dermatology at Yale University is one such modern researcher. In his treatment of a rare form of malignant skin cancer called Cutaneous T Cell Lymphoma, he refers to the photoluminescent process as "photopheresis." The important question, however, is not one of semantics, but of credit due. None of the current researchers credits earlier work in the field.

No one can deny the contributions made by Dr. Edelson to the field of phototherapy, but shouldn't the true inventors of the process be given their due? Edelson's instrument for exposing the blood to ultraviolet light is a monument to modern technology. However, it is basically a refinement of the Knott hemo-irradiator invented in the late 1930s. Edelson's instrument separates the nucleated blood cells from the red blood cells in plasma, and, according to Edelson, increases the effectiveness of ultraviolet irradiation of the blood. But the cost of treatment is increased 20 times over the Knott method, and is out of the price range of the average patient. The filters alone, which have to be replaced with each treatment, cost in the hundreds of dollars and the photoactive drug used, Oxsoralen-Ultra (8-MOP), is so expensive that it, too, is beyond the means of the average patient.

Edelson's most important contribution to phototherapy is in the field of photoactive drugs (photopharmacology), but, as Knott and the other early scientists proved, photoactive drugs are not necessary for effective therapy. These drugs and Edelson's upgraded hemo-irradiator may improve photoluminescent therapy, but the benefit-to-cost ratio is minuscule – and ultraviolet therapy as practiced by Knott was almost 100 percent effective without such modern innovations.

Despite the advanced and critically important nature of early photoluminescent research, none of the writings of present-day clinicians give even passing mention to it. This, in spite of the fact that earlier research was frequently published in the best journals of the day. Many of the references in this book are evidence of that fact.

Further, when Edelson applied for a patent on his photo-irradiation device, he was quite silent on earlier, similar patents. When applying for a patent, the applicant is required to list all similar devices previously patented. But Edelson listed a device patented by a gentleman named Edblom in September 1928, and then skipped *40 years* of patents and references. His next reference is that of "Jones patent number 3,325,641," dated June 1967. Suspiciously, this 40-year leap leaves out three of Knott's patents which strongly resemble the device Edelson designed. An inadvertent omission is certainly possible, but if Edelson could not find Knott's patents (listed under "ultraviolet" and "irradiation of blood") Edelson is not the researcher he claims to be.

Although there is nothing illegal in all of this, it is typical of modern scientists who refuse to give credit where it is due. Perhaps an overactive ego is behind such refusal; perhaps it is a desire for fame or fortune. Whatever the cause, a great injustice is being done to many noteworthy pioneers of modern medicine – pioneers who seem to have considered their hippocratic oath far more important than such things as ego or money. The discoverer of the technique of photo-irradiation of the blood, Emmett K. Knott, should have gotten a Nobel Prize. Instead, he is virtually forgotten.

Some of my advisors say I shouldn't make an issue of this and should let this book be entirely "positive." My answer is that no one can be entirely positive who accepts the right and refuses to expose the wrong (thereby accepting it, as well). In order to affirm the goodness of the good, one must contrast it with the bad. Otherwise, the good has no distinction.

I would not be worth listening to if I did not have some commitment to justice and the rights of others. Part of the overwhelmingly positive news that this book represents is that photoluminescence is a thoroughly researched, tried-and-true procedure. Its record of success is long, distinguished, and unbroken; the pioneers who developed it are eminently reliable. The reason you don't know about it is that mistakes have been made (over-reliance on antibiotics) and injustices have been perpetrated (failure to credit earlier research in the field). But such "negative vibes" accentuate the good news of this book. In fact, they were the reason I needed to write it!

If you are still not convinced, permit me to list a few other examples of misplaced recognition. A scientist named Beauchamp actually did much of the research with which Louis Pasteur was credited. The Pasteur Institute should actually be called the Beauchamp Institute. William Jenner, a quack who bought his diploma, stole the idea of smallpox vaccination from a Welsh farmer named Benjamin Jesty. Marconi did not invent the telegraph; the Wright Brothers were not the first to fly; Robert Gallo did not discover the AIDS virus; and the list goes on.

If you are going to believe in the efficacy of photoluminescent therapy, we need to right the wrongs done to earlier researchers. Only in its proper historical context will photoluminescence be recognized for its verified reliability. Knott, in particular, should be honored for discovering and perfecting one of the greatest therapeutic methods of this, or any other age.

Dr. George Miley, commenting on the pioneer work of E. K. Knott, said in 1940, "I think personally that this is one of the greatest contributions to medicine ever made by a citizen of the United States."

That statement is just as true now, more than 50 years later.

Chapter 3

Clinical, Physiological, and Theoretical Considerations

By 1944, the results were in, and it was quite clear that ultraviolet irradiation of blood was a sensationally effective means of treatment, obviating the use of antibiotics in most infections.

Dr. Henry A. Barrett reported that, in his opinion, many severe infections were efficiently controlled by the use of the Knott Photoluminescent technique. Additionally, researchers agree that the Knott technique is a consistent and reliable method of controlling acute pyogenic (pus-forming) infections (other than bacterial endocarditis – inflammation of the inner lining of the heart – which uniformly failed to be influenced by the therapy; the reasons for this singular failure of photoluminescent treatment was not known and will be discussed in a subsequent chapter).

Dr. Miley showed that ultraviolet blood irradiation therapy would quickly cure viral pneumonias.[1] He reported:

1. Complete subsidence of toxic symptoms 24 to 76 hours after a single treatment.
2. Disappearance of cough in three to seven days.
3. X-ray evidence of complete clearing of the lungs within 24 to 96 hours after a single treatment.

These results are confirmed by other researchers.[2] Dr. Miley remarked, "The relatively rapid disappearance of all

1 Miley, *American Journal of Bacteriology*, 45: 303, June 1943.
2 Hancock, *Northwest Medicine*, 33:200, June 1934.

subjective and objective signs and symptoms of pneumonia due to a virus seems significant, *as there is as yet no other therapy capable of producing the same results.*" But in spite of these findings, this remarkable therapy was discontinued and remained dormant for 50 years!

One of the most interesting things about the photoluminescent technique is that it does not kill bacteria directly. This fact is critical since only 1/25th of the total blood volume is actually irradiated in the Knott technique.[3] The procedure simply applies ultraviolet rays to the blood utilizing a "photon pump" (an ultraviolet light source) in an optimal dosage. Then, when reinjected, the irradiated blood somehow activates the rest of the body's defenses to attack the infection. Experiments have shown that bacteria mixed with blood and then directly irradiated with ultraviolet light for the amount of time used in the Knott technique has no effect whatsoever on the bacteria.[4] So the clinical effect is obviously not due to direct killing by ultraviolet rays. The cause is instead a much more subtle and complex mechanism closely connected with luminescence and "ultraviolet ray metabolism" – the actual metabolizing, or using in the body's chemical processes, of light energy – and photonic energy.[5]

Apparently, such subtlety and complexity were barriers to credibility for many in the medical profession. Dr. Barrett endeavored to explain some of the difficulty:

> To those bacteriologists and others whose experience with ultraviolet energy has been largely confined to experiments consisting in the exposure of petri dish cultures to various sources of ultraviolet radiation, it is only fair to point out something that has apparently escaped many such an investigator, namely, the vast difference between inert gelatin and living, reacting animal tissues.

The unknown quantity lies in the body, in the blood, not in dead gelatin.

> Dr. Barrett further pointed out that the remarkably positive effects ultraviolet light has on various diseases are a deep mys-

---

3 1/250th *of the blood volume* in the Douglass technique.
4 Guttmacher, and Mayer: *American Review of Tuberculosis*, 10: 170, October 1924.
5 Miley, *Archives of Physical Therapy*, Vol. 25, June 1944.

tery. Granted for sake of argument that bacteria are readily killed under certain conditions by ultraviolet energy, granted that they are also killed if present in the blood of a patient when some of this blood passes through the irradiation chamber, and that toxins are inactivated also, but how is one to explain the apparent effects on the balance of the blood not treated, i.e., 19/20 of the blood volume?

It has been suggested that the irradiated portion of the blood carries the primary ultraviolet rays into the untreated portion of the blood and that secondary radiation is produced in this way. But the author cannot entertain this theory, from the point of view of the radiologist, with a great enthusiasm. The initiation of chained photo-chemical reactions in the irradiated blood continues in the unirradiated portion and is probably the basis of the action. When we consider first of all the complexity of even the most simple photo-chemical processes and then reflect on the very complicated character of blood, consisting as it does of serum, red and white cells, platelets, chylomicrons, antigens and antibodies, hormones, endocrine substances, enzymes, iron, phosphorus, calcium, and various pathogenic substances at times, bacteria, bacterial and other toxins, and so forth, it seems somewhat foolhardy to undertake to say just what happens under the influence of ultraviolet radiation.[6]

The exact mechanism by which the small amount of irradiated blood causes its dramatic effects may never be entirely understood. The physiologic responses undoubtedly have their origin in photo-chemical reactions produced when the ultraviolet energy is absorbed. The effect is physical, then chemical, and finally, biological.[7]

Is there such a thing as an ultraviolet-light deficiency? Ultraviolet light has always been an integral part of the environment of all forms of life existing on the surface of the earth. It does not seem illogical to presume that ultraviolet rays are as much a part of our life as are optimum heat, sufficient vitamins, optimum oxygen supply, adequate diet, and other general, known entities, without which we cannot live.

---

6 Barrett, H. A., *American Journal of Surgery*, July 1943.
7 Laurens, *Physiological Effects of Radiant Energy*, P. 568.

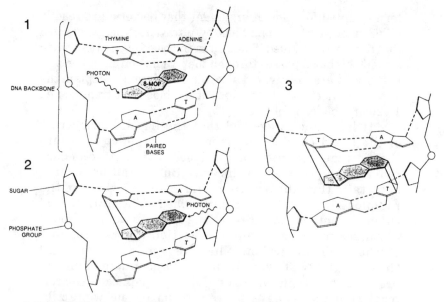

8-MOP BINDS TO DNA after being activated by ultraviolet radiation (8-methoxypsoralen). Adapted from *Edelson Scientific American,* Aug., 1988.

Although "ultraviolet-light deficiency" is not beyond reason, there seems to be more to it than that. A direct stimulative effect over and above the body's ultraviolet requirements seems to take place, causing some sort of energizing action on the immune system and perhaps other systems of the body. For whatever reason, ultraviolet light seems to act as a powerful regulatory and normalizing agent on all of the body's systems.

Dr. George Miley summarized it well:

> The expression "ultraviolet ray metabolism" implies a well-recognized physiologic phenomenon the mechanism of which is easily demonstrable and clearly understood. Actually, the subject is clouded by our ignorance of the subject. Eventually, after much long and painstaking work, we hope to find this cloud brushed away and to obtain a much clearer picture of ultraviolet ray metabolism than is possible in light of present day knowledge.

Because of the brilliant work of N. R. Finsen, we know there is indeed an ultraviolet ray metabolism.[8] Finsen discovered that

---

8 "Mitteilungen," Part One: 10, 1900.

hemoglobin absorbs all wavelengths of ultraviolet light. There can be no doubt that adequate ultraviolet ray intake is of primary importance in the maintenance of normal health. The corollary to this is that many disease states follow on the heels of a subnormal ultraviolet ray metabolism.

Dr. Gurwitsch, researching photo-biology in the 1930s, made some even more remarkable discoveries.[9] He demonstrated "mitogenic rays," tiny emanations given off by body tissues in different wavelengths, all in the ultraviolet spectrum and varying in wavelength according to the organ emanating the rays. Modern science and scientific investigators have ignored this remarkable discovery until recently.

Photonic medicine should soon be used for diagnosis as well as therapy. Photo-counting equipment used in astronomy can now be used to detect the light emanating from living cells. Living cells emit extremely low levels of visible light in the dark, as Gurwitsch proved in 1936. Specifically, they emit between 0.1 and 100,000 photons per second per square centimeter, according to researchers at Tohoku University, Japan.

Humio Inaba, one of the researchers at Tohoku, says that this photon light emission "clearly is associated with a variety of vital activities and biological processes."

It is entirely possible that we will soon be diagnosing illness by monitoring these light emissions from cells. Smokers, for instance, emit about twice as much light from their cells as non-smokers (you'll probably never think of a smoker "lighting up" in the same way again). The smokers' cells seem to be crying out for help. Within a day after cessation of smoking, the light emission level drops down to that of non-smokers. Smokers' urine also emits more light than that of non-smokers.

Inaba also found that the blood of patients suffering from diseases, such as cancer, diabetes, and jaundice, gives off much more light than the blood of healthy people. But paradoxically, although sick cells emit light, they actually have an ultraviolet deficiency.[10]

---

9 Gurwitsch, "Invisible Radiations of Organisms, Protoplasma – Monographin, Burlin, Borntraeger," 1936, Vol. 9.

10 Rahm, Otto: The Physio-Chemical Basis of Biological Irradiations, Cold Spring Harbor Symposia and Quantitative Biology, Vol. 2, 1934, pp. 226-232.

The Soviet scientist Vlail Kaznacheyev proved, over 20 years ago, that photons are emitted from diseased tissue. Almost unbelievably, he found that a disease can actually be transmitted through the air *by the photon emission* if a medium, such as radar, is used to carry it. In other words, the actual disease organism need not be transferred, only its photon emission, in order for the disease to be transferred to another individual.

Effect of Ultraviolet Light on the Blood

The number of red cells, white cells, and platelets in the blood may increase following irradiation with ultraviolet light. Such irradiation also produces a lowered blood sugar, an increased sugar tolerance, and an increased blood calcium.[11] Although blood cells can be made to increase with ultraviolet, a lowering of the white blood count can also be induced with certain frequencies of ultraviolet light.[12]

Dr. Harry Laurens of New Orleans stated in 1938:

> There is no unequivocal evidence that ultraviolet radiation increases resistance to specific or general infection, although a relationship between sunlight and the general course and character of disease, growth and nutrition has been demonstrated.

In the greater context of his work, what Dr. Laurens is saying is there's no evidence that taking ultraviolet treatments will prevent one from catching disease. However, treatment with ultraviolet light after contracting certain diseases decreases the duration and virulence of the illness, often quite dramatically.

Increase in the Efficiency of Oxygen Exchange

One of the most obvious physical reactions to photoluminescent therapy is a rapid increase in the pinkening of the skin. This reaction has been observed by virtually all researchers

---

11 The promising treatment of diabetes mellitus with UV irradiation needs further investigation.
12 Kennedy, W.P., et al, *Journal of Physiology*, 87: 336.

using the technique. The pinkening was first believed to be due entirely to peripheral vaso-dilatation (expansion of the blood vessels next to the skin), but was later ascribed to both dilatation *and* an increased ability of the blood to pick up oxygen.

In 1939, Miley made a study of the effects of 97 blood irradiation treatments given to persons suffering from various diseases. He observed:

1. A 58 percent increase in venous oxygen (oxygen in the veins) in ten minutes.
2. A nine percent decrease in venous oxygen after one-half hour.
3. A 50 percent increase in venous oxygen one hour to one month after treatment.

These changes were not due to any increased oxygen-carrying capacity of the red blood cells because there was no rise in hemoglobin or red cell count to account for the oxygen increase. The additional oxygen was therefore in the plasma itself – the non-cellular portion of the blood.

The majority of the increases in venous oxygen values occurred in persons whose values were below normal. Many of the early investigators reported quite frequently on the dramatic decrease in cyanosis (a bluish skin color caused by lack of oxygen in the blood) in patients who were near death, but who then received photoluminescent treatment.

There is undoubtedly an increase in efficiency of oxygen exchange and oxygen metabolism following the use of ultraviolet blood irradiation treatment. Unfortunately, arterial oxygen measurements were not performed extensively in 1944. Methods of checking arterial oxygen content are now much more simple and precise. It would be interesting to see the effects of photoluminescent therapy on oxygenation using such modern methods. I would expect even more dramatic findings than those documented in earlier research.

Cardio-Vascular Effects of Ultraviolet Treatment

Not enough research has been done on the long-term effects of photoluminescence on the cardio-vascular system. We don't know, for instance, what effect light therapy has on congestive heart failure. The overwhelmingly positive effects of photoluminescence certainly indicate there may be long-term benefits for such things as hypertension and heart failure, however. Russian doctors are currently working in this clinical area. (See Chapter 21.)

Effect on the Autonomic Nervous System[13]

The peripheral vaso-dilatation mentioned above requires further attention – particularly insofar as it parallels many other effects on the autonomic nervous system.

Within a few minutes, approximately 75 percent of those receiving ultraviolet blood irradiation will have this dramatic "pinking up." It may persist for more than 30 days, and it is usually accompanied by a feeling of well-being. Its appearance is almost always regarded as a favorable sign in whatever disease is being treated.

This remarkable effect on the autonomic nervous system is also illustrated in the rapidity with which paralytic ileus (a paralysis of the bowel) can be relieved with ultraviolet blood irradiation.[14] Researchers consistently note that the smooth muscle tone and ability to contract of the gastrointestinal musculature are restored to normal 12 to 24 hours after a single blood irradiation. These effects obviously indicate that the portion of the autonomic nervous system controlling the gastrointestinal tract returns to normal following photoluminescent therapy.

Part of the favorable response of thrombophlebitis (infection and inflammation of superficial or deep veins) to photoluminescence can be attributed to this effect on the autonomic nervous system. The response is so dramatic it's hard to believe it can be due to destruction of the bacterial infection alone.

Allergic asthma also responds dramatically to photoluminescence. Its response may similarly be attributable to the effect of treatment on the autonomic nervous system – specifically with regard to the smooth muscle of the bronchial tubes. Many of the cases treated by various investigators were considered to be extremely difficult to treat. Such intractability is

---

13 The portion of the nervous system that controls involuntary action, such as that which occurs in the intestines, heart and glands.

14 Rebbeck, *Review of Gastroenterology*, January-February, 1943.

not encountered with photoluminescence, however, lending credence to the belief that photoluminescence relaxes the bronchial tubes indirectly through its effects on the autonomic nervous system. Whatever the exact cause, the end result is dramatic relief for the patient. In fact, about 80 percent of asthma patients respond favorably to photoluminescence, as observed by the relaxation of the smooth muscle in their bronchial tubes. The implication is that the autonomic nervous system, being hopelessly out of balance, is restored to normal by photoluminescent therapy.

Metabolic Effects

Ultraviolet irradiation typically causes the body to rid itself of uric acid more rapidly, which suggests possible usefulness in the treatment of gout (gout is an arthritic condition caused by the deposition of uric acid in the joints). Irradiation also doubles the fat content of the blood, and the cholesterol may increase by as much as 30 percent. This should not necessarily be considered a negative response and may, in fact, be very positive. Cholesterol is *vital* to your body's immune defenses and is not the terrible poison that the food industry would have you believe.

Researchers have also found that blood sugar is temporarily diminished in diabetics by ultraviolet irradiation. This effect is probably due to an increase in the excretion of insulin. Other research has found that ultraviolet irradiation prevents and cures rickets, but is ineffective in healing fractures.

Tetany, a low calcium state which causes spasm of muscles, can be relieved with ultraviolet irradiation, though such is not the treatment of choice. The therapeutic effect is so strong that even an animal in which the parathyroid glands are removed (thus preventing the absorption of calcium) can be prevented from having tetany through photoluminescence. However, vitamin D is even more efficient in preventing tetany.

Effects of Physiologic Conditions on Sensitivity to Ultraviolet Light

One of the results of sensitivity to ultraviolet light is the tendency to get a sunburn with minimal exposure. Research has shown that varying levels of sensitivity to ultraviolet light from person to person are normal, and we've all known those who seem to burn easily in the sun. However, persons with unstable nervous systems, overactive thyroid glands, elevated blood pressure or active tuberculosis have shown significantly increased sensitivity to ultraviolet light. Many drugs, including antibiotics, also have a tendency to dramatically increase ultraviolet sensitivity. Another factor is the menstrual cycle, which increases sensitivity the greatest amount on the first day. Pregnancy brings an increased sensitivity, as does an acid diet, but alkaline salves increase skin sensitivity more than acid salves.

The Possible Effects of Ultraviolet Light on the Pineal Gland

I was taught in medical school (in the mid 1950s) that the pineal gland, located at the bottom of the brain, was some sort of leftover from "evolution."[15] The standard joke was that "it's the seat of the soul." My *Dorland Medical Dictionary*, 20th edition, describes the pineal gland as a pine cone-shaped "body" located at the epithalamus and an outgrowth thereof. "It is not composed of nervous [sic] elements and is a rudimentary [undeveloped] glandular structure said to produce an internal secretion."

Since my time in medical school, the pineal *body* has become the pineal *gland*. It is the body's chief photoreceptor – a receptor of photons from the eyes – and thus protects the body from light deficiency.

Calcification of the pineal occurs in a large percentage of people over 60 years of age. It is not a "natural part of the aging process" because it doesn't happen in everyone, but it is very common.

As an emergency medicine specialist for 20 years, I saw many sick elderly people. The calcified pineal gland was an almost routine finding, and still is. It's so common that it is used as a diagnostic tool for finding a space-occupying mass in the brain. If an x-ray reveals that the pineal has shifted to one side from its normal place in the mid-line, the presence of a mass which is pushing it off center can be ascertained. Obviously,

---

15 A theory at serious odds with the fossil record and many laws of science at the present time – though scientific implausibility seems not to deter its adherents. If you cite Darwin often enough and with sufficient authority; if you use only incomprehensible, esoteric, twenty-letter words; and if you add another ten billion years to the age of the earth each time you make a discovery that doesn't fit your preconceived notion of origins; you can still travel comfortably in the highest scientific circles anywhere in the world.

this technique is of no help in diagnosing healthy elderly people, because you can't see tissue on an x-ray if it is not calcified.

If a method could be devised to decalcify the pineal gland, I believe wondrous health benefits would ensue. Some of the benefits of chelation therapy using EDTA and/or photo-oxidation may be due to partial decalcification of the pineal gland.[16] A series of CAT scans of the pineal gland, before and after chelation, should resolve the issue.

For further information on the chelating qualities of the oxygenation process, see my book *Hydrogen Peroxide: Medical Miracle*.

Effect of Ultraviolet on the Skin

The skin is one of the largest and most important organs of the body. One of its many functions is to absorb sunlight for the prevention of rickets through the utilization of vitamin D. This process occurs in a layer of the skin called the horny layer. Inflammation of the skin caused by sunburn takes place in the basal cells of the malpighian layer, which is a little deeper.

The horny, clear, and granular layers of the skin act as light filters, keeping much of the surface light from reaching lower skin layers. Those ultraviolet rays which do pass through activate the ergosterol (a hormone) and cholesterol located in the inner skin layers. The ergosterol, when irradiated, is converted to vitamin D.

What we refer to as sunburn is actually an inflammation of skin cells in reaction to excessive sunlight. The skin's tendency to burn varies with an individual's complexion, sex, and age. Blondes are from 40 to 170 percent more sensitive than brunettes, men 20 percent more sensitive than women, and persons between twenty and fifty are more sensitive than those younger or older. Maximum sensitivity occurs in March and April and in October and November.

---

16 Chelation therapy is a method of removing harmful metals from the body. The chemical EDTA is a powerful chelator – it combines with toxic metals and brings them into solution so the kidneys can excrete them through the urine. Assuming that calcification of the pineal gland is detrimental – which is a reasonable assumption – then any method of chelation which includes EDTA and photo-oxidation would greatly benefit the patient's light physiology, since the pineal gland is the center of light metabolism for the body.

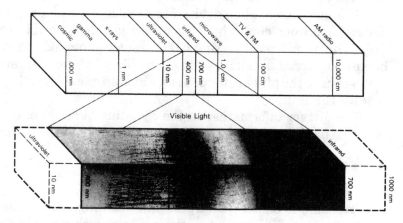

Electromagnetic Spectrum

Photoluminescent therapy may have application for treatment of disease in all wavelengths of light from ultraviolet through infrared.

The Parameters of Therapy

Photoluminescent therapy involves four parameters:

1. The amount of blood taken from the patient.
2. The time of exposure to the light.
3. The intensity and the wavelength of the spectral energy used.
4. The sensitivity of the photo-active drug used (if a drug is used).

Fundamental Laws

The fundamental physical laws involved in the mechanism of photoluminescence are:

1. The Grotthus-Draper Law, which is known as the first law of photo-chemistry: Only light which is absorbed can act chemically.
2. The Bunsen-Roscoe Law: A photo-chemical change is proportional to the intensity and the time of illumination of the tissue.

Chapter 4

The Detoxifying Effect of Ultraviolet Blood Irradiation

One of the many useful aspects of light is its deadly effect on toxins. Toxins are poisonous substances, such as animal (snake or scorpion venom), or bacterial (strep, botulin, tetanus, etc.) poisons. Exposure of such toxins to light quickly deactivates them. The potential health benefits from this phenomenon are incalculable.

This anti-toxin effect of light has been known for over 100 years.[1] Early research, done in the 1800s, paved the way for critically important early 20th century work in the treatment of such toxins as tetanus and botulism. So totally does light deactivate toxins that no one need die of rattlesnake bite, tetanus, botulism, or rabies anymore, if phototherapy is close at hand.

Early research with snake venoms concentrated on the effects of sunlight, rather than of ultraviolet blood irradiation, on toxins. Obviously, such research has application to later work in photoluminescence.

A report from the Rockefeller Institute for Medical Research in about 1905[2] described the use of a certain dye in conjunction with direct sunlight and their combined effect on various snake venoms. The researchers exposed crotalus venom (that found in

---

1 "Scientific Memoirs by the Medical Officers of the Army of India," 1895, Part IX, and 1898, Part XI.
2 The copy of the paper from which this material was taken is undated, and the journal is not identified.

rattlesnakes, water moccasins, and copperheads) to sunlight in the presence of the dye. They found that the venom was reduced in strength by almost 200 times when measured by the absence of hemolysis (destruction of red blood cells). Without the sunlight, however, the dye had no effect on the venom. The venom would destroy all red blood cells in the dark, even though the dye was present.

The technique used by Dr. Hideyog Noguchi at the Rockefeller Institute was quite simple. Stock solutions of snake venom were mixed with dye. The mixture was then divided into two parts, one of which was exposed to sunlight and the other kept in the dark. The exposure to sunlight was of 30 hours duration, as a rule.

Noguchi then took the venom, both exposed and unexposed, and injected it into rabbits. When even low doses of unexposed venom were injected, the rabbits quickly died. However, a low dose of venom that had been exposed to sunlight would only make the animal slightly ill, following which it would recover completely.

In each of these experiments, eosin, a photo-active dye, was used for photo-dynamic effect. But during this research, it was found that the anti-toxin effect of eosin and sunlight on cobra venom was minimal. This fact is probably due to cobra venom being primarily a neuro-toxin (a toxin which attacks nerves).

Crotalus venom, such as found in rattlesnakes, contains a preponderance of a toxin called "hemorrhagin," which destroys red blood cells. Crotalus venom also contains a small quantity of neurotoxin, but it is not of primary effect in this type of venomous snake.

Noguchi used guinea pigs for the testing of crotalus, with the treated or untreated toxin being injected into the abdomen. His studies proved that the crotalus venom, when treated with a photo-sensitive compound and then exposed to sunlight, would not kill guinea pigs. Thus, the 30-hour exposure period, though insufficient to deactivate cobra venom, was more than sufficient to radically reduce the toxicity of rattlesnake venom.

Noguchi also found that the intravascular clotting factors (elements in the blood causing it to coagulate) were completely neutralized in rattlesnake venom when it was treated with a photo-sensitive compound and exposed to sunlight.

* * * * *

Though incidences of botulism (a deadly form of food poisoning) have declined with modern advances in food preparation and refrigeration, most people are still aware of how serious it is. Every now and then someone gets poisoned by some food prone to producing botulinum bacteria and becomes violently ill. Every time it happens, we are reminded of the extreme toxicity of this form of bacteria.

However, photoluminescence is very effective in treating botulism. One of the most easily demonstrable clinical effects of blood irradiation is the alleviation of toxic symptoms within 24 to 72 hours after treatment. Miley reported a most remarkable case in which this effect was strikingly illustrated:[3]

The patient was unable to swallow or see at the time of irradiation. Within 48 hours, the patient, who obviously was dying from the well-known effects of the classic botulinum neurotoxin, was able to swallow and see and was completely clear mentally. Thirteen days after *a single blood irradiation treatment*, she left the hospital in excellent condition and remained so. As Miley said, *"There is to my knowledge no record in medical science of any other therapy that can produce such an effect on a patient in the last stage of botulism."* It is depressing to contemplate the number of people who have died from botulism, tetanus, rabies, gas gangrene, snakebite, and other toxins because of the abandonment of this incredibly safe and effective detoxifying therapy.

Many bacteria, in fact, perhaps all, kill by producing a toxin. The Staphylococcus, the typhoid germ, tetanus, gas gangrene, and the organism of plague, Ursinia pestis, all kill at least partially through the effect of very potent toxins produced by the organisms.

Toxin neutralization is quite a remarkable and mysterious action of ultraviolet light therapy. For instance, patients with tetanus, botulism, and snake venom, both the hemolytic (blood destroying, such as rattlesnake venom) and, to some extent, neurotoxic (nerve-destroying, such as cobra venom) varieties can be neutralized by photoluminescent treatment of the blood. Even the most deadly of snakes, such as the fer-de-lance and the deadly

---

3 Miley, *Archives of Physical Therapy*, Volume 25, June 1944.

tropical rattlesnake (whose bite is nearly 100 percent fatal), could be neutralized by ultraviolet therapy if administered in time. But these more venomous snakes are what are known as "two steppers." Treatment with ultraviolet light is difficult when the victims are dead after taking two steps.

* * * * *

Another property of ultraviolet treatment, which could be of tremendous importance in time of war, is its effectiveness against biological warfare toxins. Ricin, an extract from the castor bean, is a fairly common biochemical warfare agent. Rapid administration of ultraviolet light causes the ricin to lose its hemolytic (blood destroying) power.[4] Since recovery after blood irradiation is usually rapid and complete, the advantages of having this technology could be decisive for the side possessing it. This is especially true for the United States, since the U.S. has not aggressively pursued development of biological agents (and therefore may not always know how to effectively counteract them), but faces enemies who have highly effective biological weapons.

Another germ which kills by toxin is diphtheria. Dr. Yersin, the discoverer of the plague organism, found that the diphtheria toxin is completely inactivated by simply exposing it to direct sunlight.[5] Italian doctors, as far back as 1891, found that tetanus toxin is also sensitive to sunlight.

Soon after that finding by the Italians, Doctors Flexner and Noguchi also noted the photo-dynamic destruction of tetanus toxin by placing it in a solution containing eosin (the photo-sensitive dye mentioned earlier) and exposing it to sunlight. This was probably the first experiment combining ultraviolet light with a photosensitive drug – a technology now being investigated all over the world.

* * * * *

An intriguing article by Dr. D. I. Macht, entitled "The Influence of Ultraviolet Irradiation on Menotoxin and Pernicious

---

4 Jodlbauer and Tappeiner, Leipzig, Monograph 1907 referenced in *Archives of Physical Therapy*, September 1942; Carmichael, *Journal of Pharmacology*, 35: 193, 1929.

5 *Annals of the Pasteur Institute*, 3: 273, 1889

Anemia Toxin," was reported in 1927.[6] This paper was written before pernicious anemia was found to be a vitamin deficiency (there is no known "toxin" involved in pernicious anemia; it is a severe anemia caused by a deficiency of vitamin B_{12}). It is interesting to speculate on what effect the ultraviolet light had which would cause the remission of a vitamin deficiency. Is ultraviolet light involved in B_{12} metabolism, as well as in the metabolism of Vitamin D? No answer to that question is yet known.

Dr. George Miley made the important observation that bacterial toxins are neutralized more readily by the presence of oxygen. This raised an interesting possibility. Perhaps the effectiveness of ultraviolet irradiation therapy could be increased by giving it concomitantly with intravenous hydrogen peroxide.[7] Since hydrogen peroxide breaks down in the blood to water and oxygen, more oxygen is available to assist in the neutralization of toxins. I introduced this concept in a clinic I run in Africa and have found it to be superior to using photoluminescence alone. I call the combination "photo-oxidation therapy."[8]

Dr. Miley was profoundly impressed by the detoxifying effect of the ultraviolet therapy and devised what he termed a "detoxification time" for various maladies he and others had treated. This detox time has two components, which he called "early" and "complete."

In patients with a generalized peritonitis (inflammation of the lining of the intestine, often fatal in Dr. Miley's day and still occasionally so today), 100 percent of those with a moderately advanced peritonitis, and 53 percent of those who were apparently doomed to die, recovered from their illness. The early stage of detoxification was accomplished in an average of 34.5 hours, with complete detoxification (recovery) in an average of 81.75 hours.

Those patients with an abscess from appendicitis had a somewhat more rapid recovery. The treatment was so effective that 75 percent of those who were expected to die (those not responding to antibiotics and surgery) survived. The early

---

6 *Proc. of Soc. Exper. Biol. and Med.*, 24: 966, 1927
7 *Archives of Physical Therapy*, September 1942.
8 See Chapter 15 on Africa.

detoxification time was 24 hours, and recovery was complete in 53 hours – a little over two days.

The most impressive early sign of recovery, Dr. Miley noted, was the dramatic and precipitous fall in temperature.

Dr. Miley reported on the remarkable uptake of oxygen in venous blood noted after ultraviolet blood irradiation therapy. This rise in oxygen content of the venous blood is marked and persistent in patients who are found originally to have a low venous oxygen level. Miley and the other investigators did not check arterial oxygen because procedures to do so were not generally in use in the early 1940s.

The rise in oxygen was not accompanied by a rise in the blood hemoglobin or red cell count, which means that the plasma of the blood was carrying a higher content of oxygen outside of the red cells. The mechanism of this increased oxygen content of plasma is not understood. (Other researchers have reported a rise in the number of red cells following ultraviolet therapy, and they report that the cells have increased oxygen-carrying capacity.) The detoxifying effect may be due to the high oxygen levels obtained.

An interesting example of this phenomenon is the case of a young woman admitted to Hahnemann Hospital in Philadelphia. She was in extremely serious condition with blood poisoning from a criminal abortion. A full week of intensive antibiotic therapy had failed to control the progress of her infection. At the time she was seen for blood irradiation, she was extremely debilitated, toxic, jaundiced, and near death. On the first day following irradiation of her blood, her toxic symptoms began to subside. Forty-eight hours after irradiation therapy, she was out of danger and her detoxification was complete. This case and the two following cases are remarkable in that *only one treatment* completely reversed an apparently hopeless situation.

A patient contracted peritonitis following a Caesarean section. After 24 hours of intensive antibiotic therapy, it was noted that the blood count had begun to fall and the temperature to rise with an associated rapid deterioration of her general condition. Blood irradiation was instituted when the temperature reached 106.8°F. Within the next 24 hours the patient's symptoms subsided markedly and the detoxification was complete in 24 hours. The patient convalesced with no further problems.

Ultraviolet irradiation was clearly lifesaving in a case of acute gangrenous appendicitis with generalized peritonitis. At the time of initiating blood irradiation, the patient was in profound shock from an overwhelming toxemia and general peritonitis. This condition arose from the removal of his gangrenous appendix 48 hours previously.

Despite the fact that the patient was in shock, 150 cc of blood was withdrawn, irradiated, and quickly given back to him. Within a few minutes, his pulse began to improve markedly; the signs of shock rapidly disappeared. On the first day following his blood irradiation he could sit up in bed and write a letter. His rapid recovery from a toxic near-death condition was astounding.

The following case illustrates the usefulness of ultraviolet blood irradiation as a prophylaxis (preventive treatment) prior to surgery:

A patient suffered from multiple pelvic abscesses and pus in her fallopian tubes. This condition caused her to steadily deteriorate, despite antibiotic therapy. On the 14th day of her hospital admission blood irradiation was performed, which resulted in complete detoxification within 72 hours. Her temperature remained elevated, however, due to the presence of accumulated pus pockets in her pelvis. So a prophylactic ultraviolet blood therapy was given prior to surgery to neutralize the pus. The operation was performed, successfully evacuating the large area of infection. Her temperature fell rapidly and she recovered with no side effects whatsoever. She was discharged, completely well, within 24 days of surgery.

* * * * *

A young male patient was close to death from generalized peritonitis arising from a gangrenous appendix. He also had complete paralysis of the bowel and secondary pneumonia with severe difficulty breathing and cyanosis. Death was expected within 24 hours.

In the first few minutes following the initial blood irradiation, the patient's arrhythmia (irregularity of heartbeat) and cyanosis dramatically disappeared – clearly a detoxifying effect. However, two more irradiations were required before the patient's toxic symptoms completely subsided. The patient continued to have an elevated temperature for some time. But after his abdominal drain was removed, and without any further ultraviolet light treatment, his temperature rapidly returned to normal. He was discharged entirely well 33 days after his initial irradiation.

* * * * *

The detoxifying action of ultraviolet light makes it useful in the preparation of various vaccines. The fundamental theory is that ultraviolet irradiation will often inactivate the toxic properties of bacteria and viruses, even snake venom, without destroying the ability of the toxin to act as a vaccine protective agent in humans and animals. Dr. V. L. Trotskii noted in 1936 that bacteria killed by ultraviolet light lose a considerable portion of their toxicity toward test animals but completely retain their antigenic (ability to cause the body to produce antibodies) and immunizing characteristics. This principle has been used for many years in the manufacture of various vaccines. Doctors Smithburn and Lavin demonstrated that the tuberculosis bacillus, when irradiated with ultraviolet light (2,537 angstrom units) is rendered non-toxic, yet it is still viable and so possesses the capacity of producing considerable immunity without causing disease. The rabies virus[9] reacts in a similar fashion, as does the eastern equine encephalitis virus.[10]

---

9 Hodes, *J. Expr. Med.*, 72: 437, 1940.
10 Morgan, *Proc. Soc. Extr. Biol. and Med.*, 47: 497, 1941.

Chapter 5

Beyond Antibiotics: Photosensitizers

As effective as Knott's technique of photoluminescence was, other means have been found to make light treatment even more effective. I mentioned briefly my technique of using photoluminescence in concert with hydrogen peroxide. Others, however, have used agents called photosensitizers[1] – chemicals that increase the body's sensitivity to light and thereby amplify the effects of light treatment.

For instance, researching early in this century, a scientist named Raab exposed protozoa to a toxic compound called acridine. He found that they were not affected by the toxin in a solution of 1:20,000 if they were kept in the dark. However, if they were exposed to direct sunlight while in this toxic solution, they would die in six minutes. This work was one of the first discoveries that certain compounds could kill living organisms in the presence of sunlight. Raab discovered that the increased action of light depended upon the fluorescence of certain chemical agents. He also made the remarkable discovery that when sunlight passes through a fluorescent solution, the light is robbed of its power to set up fluorescence in a second solution placed behind it. Also, the second solution will not produce a toxic action on organisms because of this strange filtering out of some vital

---

1 Any medical book worth its salt is going to have to be a little technical at some point. This chapter is that point for this book. Bear with me. I claim only that this chapter makes the book more complete, not more readable. Those who need to understand the mechanics of photoluminescent therapy will want to go into this material in some depth. Those who care only that the procedure works may want to skim over it more quickly.

element. Something is removed from the sunlight, although the light will still look "complete," i.e., white. Has the light been made photon-deficient?

Dreyer made an important contribution when he discovered that Finsen's light (ultraviolet) alone would kill the bacteria, Infusorian Nassula, in nine minutes. When combined with a photo-active compound, however, the Finsen light would kill the Nassula in only 10 seconds. This was clear proof that there was a place for photo-dynamic compounds in the treatment of infections.

Old German medical literature from the late 19th century reveals that diphtheria toxin, tetanus toxin, and ricin (a chemical warfare toxin) are all neutralized by the combination of photo-sensitive drugs and light.

Writing in the 1800s, and known as the father of photo-active compounds, a French researcher named Tappeiner found that a variety of parasitic and neoplastic (tumorous) conditions of the skin could be treated by applying the photo-sensitive compound eosin, and then exposing the skin to sunlight. He also noted that artificial light could be substituted for the sunlight.

Knott's work was taken a giant step forward when the (re)discovery was made that certain photo-sensitive chemicals can remarkably increase the effectiveness of treatment of sick cells by ultraviolet light.

The list of chemicals and compounds that can be made photoactive is almost endless. Steroids (cortisone) have been used with great success as photoactive compounds when they are combined with diazo or azide chemical groups. These chemical groups have a high degree of intrinsic photoactivity. That activity is retained when they are incorporated into another chemical, such as cortisone, which renders the entire compound photoactive. Sixteen-Diazocortisone is probably the preferred photoactive steroid because it retains a high degree of cortisone's original pharmacological activity. This photo-active cortisone, upon addition to the blood, readily enters lymphocytes, or other nucleated cells, and associates itself with the cortisone receptor sites in those cells. The cell is then photosensitized because of the diazo molecule attached to the cortisone, and will be strongly affected by exposure to ultraviolet light.

In the body, the most rapidly dividing cells (such as cancer cells) are more strongly affected by photon energy. These cells require cortisone to function and have cortisone receptors on their surface to "grab" what they need out of the blood. However, the modified cortisone containing the diazo chemical is grabbed just as readily. These molecules of modified cortisone both block the cell's ability to absorb normal cortisone and prevent the cell from utilizing its own cortisone. As a consequence, the receptor's ability to transmit cortisone, vital to the continued metabolic activity of the cell, is blocked. The cells quickly become unable to function because of "cortisone deficiency" and are rapidly destroyed.

It is possible to sensitize blood cells just as you would a photographic plate. There are many sensitizing substances, most of which are fluorescent, but the fluorescence is not the cause of the reaction. The photo-dynamic effects occur only in the presence of oxygen.[2]

Among photo-dynamic sensitizers are erythrosin, rose bengal, rhodamine, anthracene derivatives, acridine dyes, some amino-acids, methylene blue, quinine, adriamycin, rubidazone, sulfonamides, phenothiazine, tetracyclines, coal tar derivatives, chlorophyll, hypericin, porphyrins, and 8-MOP. The latter, eight-methoxypsoralen, is being used in research at Yale University Medical School. I suspect that some of the older compounds listed above are as effective as 8-MOP, but they have no appeal to drug companies. They are not patentable and therefore would not be profitable to market.

Again, I must emphasize that ultraviolet irradiation of the blood works *very well* in the treatment of many infections without the use of photosensitive drugs. These agents do greatly enhance the effectiveness of the therapy, however.

Not only blood cells can be sensitized. Antibodies can also be made photoactive. Such antibodies have great promise in the treatment of "auto-immune" diseases, such as arthritis, lupus, etc. Platinum anti-cancer drugs, porphyrin derivatives, psoralens, and some amino acids are all photoactive and are therefore candidates for photochemotherapy.

The field of photochemotherapy is expected to become so sophisticated that a particular type of abnormal cells in the body, such as altered T-cells, will be specifically targeted for a photo-cytotoxic

---

2 That's why it is important to oxygenate the blood with hydrogen peroxide intravenously before photoluminescent therapy.

dye (a dye which, when subjected to light, becomes toxic to the cell), such as rhodamine or fluorescein. The dye will be delivered to the targeted cells, then irradiated with ultraviolet light, thus wiping out only the affected abnormal cells. Even insulin can be used to link with a photo-cytotoxic dye for delivery to a targeted cell type. These "photo-cytotoxins" can revolutionize medical therapy if they are implemented on a large scale. They eliminate the need for present-day toxic drugs which affect every cell in the body, both healthy and sick. The concept of targeting only sick cells by attaching dyes to cortisone, insulin, or other agents, will enable us to approach the desired specificity needed to attack cancer and other diseased cells in the body. This therapy is already available, at low cost and with complete safety, using the modified Knott Technique (photo-oxidation).

Liposomes, spherical little fat vesicles that travel readily through the body, can also be loaded with effective amounts of cell-photosensitizing chemicals, such as the dyes mentioned, steroids, etc., to attack specified sick cells. The liposomes, coupled with photoactive compounds, are absorbed by targeted lymphocytes. The liposomes, when absorbed by the cell, are lysed (destroyed) by enzymes within the cell. Thus, the encapsulated photosensitizing chemicals are released to kill the cell. In effect, the cell "pulls the pin" on the liposome hand grenade and destroys itself.

Antibodies made photoactive show great promise in treating many forms of disease, especially cancer, because of their high specificity. When a photoactivated antibody is specific to a malignant lymphocyte cell, the irradiated blood containing the photoactive antibody will tag the sick (i.e., cancerous) lymphocytes with the activated antibody. The sick lymphocytes will thus be labeled as "enemy" and removed from the bloodstream.

The photodynamic process involved here is important and worth repeating: Cancer cells have a tendency to absorb larger amounts of photosensitive chemicals than normal cells. These photosensitive chemicals will then cause the death of the cells when exposed to certain light frequencies because the sick cells contain more of the chemical than normal cells. Although normal cells contain a certain amount of the photosensitive material, the amount is considerably less than the tumor cells absorb and the healthy cells are not materially affected.

Although much work has been done by *externally* irradiating the entire body with ultraviolet light following ingestion of a photosensitive drug (primarily in the treatment of psoriasis), the Knott method, as now modified by Edelson and others, has far greater therapeutic potential.

The original work of Finsen, in which he cured cutaneous (skin) tuberculosis (lupus vulgaris), was rediscovered when doctors in London, 30 years ago, found that jaundice in premature babies could be successfully treated by irradiating the infant with blue light. This "jaundice of the newborn" is caused by a retention of an excessive amount of pigment called bilirubin. Bilirubin derives from the breakdown of the oxygen-carrying chemical, hemoglobin, and is toxic to nerve cells. If it is present in excessive amounts for too long, it can cause brain damage. The toxic bilirubin is soluble in fat but not water. Therefore, it cannot be excreted from the liver, which metabolizes only water-soluble chemicals. Bilirubin is ordinarily made water-soluble by the baby's enzymes, but the premature baby may not have enough of the right enzyme in the liver to convert the bilirubin to a water-soluble form.

Researchers found that when the baby is irradiated with blue light, the bilirubin is changed into compounds that are soluble in water and can be readily excreted by the liver. This discovery has saved many premature babies and has prevented a great deal of brain damage in children.

The photosensitive drugs were seen as an opportunity to enhance the effectiveness of ultraviolet light on cancer cells. As cancer cells are more likely to absorb these photosensitive materials, and ultraviolet light is known to kill cells containing the photosensitive chemical, it was reasoned that if you can selectively get the photosensitive chemical into the cancer cells, and then irradiate the blood with ultraviolet light, you will selectively kill the cancer. Indeed, it has turned out to be true.

The first group of chemicals studied for this effect were the porphyrins. The porphyrins are interesting flat molecules with a hole in the middle – a molecular doughnut. They are intimately involved in all of life's processes. Without the porphyrin ring, life would be impossible. Plants require chlorophyll (magnesium porphyrin) for the absorption of light and for photosynthesis – the process by which chlorophyll-containing cells in green plants use the energy of light to synthesize carbohydrates from carbon dioxide and water. No magnesium porphyrin ring – no chlorophyll; no chlorophyll – no plants; no plants – no cows, sheep, or people. If there were only magnesium porphyrin and no iron porphyrin (hemoglobin), the world would be filled with plants but there wouldn't be any animals. So porphyrins, like oxygen and water, are an integral part of the lives of all living things. The simplest of the porphyrins looks like this:

: Porphyrin

The center of this molecule is a hole surrounded by nitrogen atoms (N). This hole is 0.2 nanometers in radius. 0.2 nanometers is 2/10ths of a billionth of a meter, a very small space. But this hole is just big enough to accommodate a number of different metal atoms. If you place an iron atom in the middle, the original purple porphyrin becomes red. If you insert a magnesium atom, it becomes green; a copper atom makes it blue.

Iron inserted in the hole of the porphyrin produces a heme, the building block of hemoglobin, which transports oxygen to the cells of the body. Iron porphyrin is also found in myoglobin (the type of hemoglobin located in muscle fibers) and the cytochrome system (proteins used in cell metabolism) of the body:

Iron : hemoglobin

The "metalloporphyrin" produced by inserting magnesium into the hole gives us chlorophyll. Chlorophyll allows plant cells to absorb oxygen, without which man would not be supplied with oxygen in nature. The chlorophyll from the magnesium porphyrin collects light from the sun (photons) and converts it into chemical energy in the plant:

Magnesium : chlorophyll

I used to say that copper porphyrins do not have any known biological use, though they are used in industry as pigments. But just when you think you know everything, new information comes along. Scientists report that the horseshoe crab (I *hate* those things – stepped on one once), which has blue blood, uses a copper hemoglobin:

Copper : paint pigment
("no biological effect")

Some people manufacture their own natural form of photo-sensitive porphyrin, which leads to a disease called porphyria. Because of a biochemical deficiency in the manufacture of the porphyrin compound in the blood (heme), extreme photosensitivity develops, leading to porphyria. This porphyrin, which lacks its central iron molecule, courses through the blood and reacts with the ultraviolet wavelengths of sunlight at the superficial skin layers.

The diagnosis of porphyria can be made simply by taking the patient's urine and placing it in the sun. There is a dramatic color change of the urine caused by the sun's ultraviolet light acting on the iron-free porphyrin. The urine turns the color of Coca-Cola.

The iron-free porphyrin, called proporphyrin, causes great tissue destruction and serious illness. However, this property of proporphyrin can be used to our advantage when it is properly administered for phototherapy. The porphyrins are still being studied intensively for possible medical use in conjunction with ultraviolet or infrared light therapy.

A courageous physician by the name of Meyer Betz did an experiment in 1913 which was one of the initial discoveries in modern photo-biological treatment. He injected himself with 200 milligrams of a material called hematoporphyrin and then observed his reaction to exposure to sunlight. He found that he became extremely sensitive to sunlight and quickly developed swelling and reddening in areas exposed to the sun. This was the first modern demonstration that porphyrins do indeed have "photo-dynamic effect" in humans.

It was many years before another giant step was taken in photo-biological therapy. In 1961, Lipson and Baldes of the Mayo Clinic demonstrated that tumors are very sensitive to porphyrins and will absorb them more selectively than normal cells. In their experiments they used something called hematoporphyrin derivative (HpD), produced by adding sulfuric and acetic acids to hematoporphyrin. This process produced a mixture of about ten compounds, some of which were photosensitive. The photo-sensitive HpD's emit a bright red fluorescence, which provides a means of detecting small amounts of cancer tissue in beds of normal tissue. They were used in this way for a number of years before researchers realized that if radiation with ultraviolet light was employed, a photodynamic effect would come into play and the tumor cells could be selectively destroyed.

It was another ten years before someone got the idea of combining these two porphyrin properties – the photo-destruction of tumor cells and the propensity to accumulate in tumor tissue. In 1972, Dr. Ivan Diamond demonstrated in California that continuing irradiation of tumor cells with ultraviolet light does indeed kill them. Then, a group of physicians from St. Mary's Hospital Medical School in London, working with Dr. Thomas Dougherty in Buffalo, New York, showed that you could use the HpD compound with light and selectively destroy cancer tissue in the body.

By 1976 there was enough knowledge accumulated to try this treatment in humans. The results were highly encouraging and research continued on a larger scale. Research groups are now doing work on tumor photo-therapy in Australia, Canada, China, Britain, Italy, Japan, Norway, and the U.S.

This variation in the ongoing light therapy revolution is called photo-dynamic therapy (PDT). PDT specifically employs the photo-dynamic action of red light on cancer cells which have absorbed hematoporphyrin derivative, HpD.

After HpD is injected intravenously, the cancer cells can be located – in the lung, for instance – by taking a tube called a bronchoscope and peering down into the lung passages with ultraviolet light. The tumor cells will then be seen to glow. A fine quartz fiber is inserted through a channel in the bronchoscope which conducts red light down to the site of the glowing cancer. The HpD in the tumor cells responds to the red light, forming "singlet oxygen" (single oxygen atoms – O – as opposed to oxygen molecules – O_2), which injures cell membranes, and mitochondrial DNA. Cross-linking of the DNA (and probably other factors) leads to the death of the cancer cells within two days, while normal bronchial tissues remain unharmed. With the cross-linking of DNA, there is an alteration of the cell so that it cannot multiply. If you stop the cancer cells from multiplying, they eventually die of old age and are not replaced by new ones. Again, the process of killing only specified cells is possible because the tumor cells are slow in eliminating the injected dye, whereas healthy cells quickly remove it from the body.

This treatment, although dazzling in its technology, is incomplete because it assumes that biology works along linear lines, whereas biology is actually very non-linear. Cancer is not a localized disease, restricted to areas that can be seen either directly or through illumination. Eliminating cancerous tissue, even with such precision as this new therapy makes possible, is not going to eliminate the hidden portion of the disease. As has been shown with breast cancer, removing the obvious malignancy does not cure the patient of cancer. Relapse can occur five, ten, or fifteen years after surgery. Early detection, no matter how early, does not prolong the life of the patient. There is simply an *appearance* of increased longevity because of the earlier diagnosis.

Biology is very chaotic, so the silver bullet approach simply is not going to work in lung cancer, for instance, any better than it has in breast cancer. The lung cancer detectives keep diagnosing at an earlier and earlier stage, which is supposed to increase life expectancy, but it has *only increased the time of awareness of the diagnosis*. Lung and breast cancer still kill people as frequently and as quickly as they have in the past.

A good example of this non-linear quality of biology is an illustration given to me by Dr. P. S. Callahan:

> The Congress of the United States is made up of some morons, some psychopaths, many selfish opportunists, a liberal

sprinkling of crooks and political prostitutes, and a small number of dedicated, patriotic statesmen. This disparate group has been meeting now for over 200 years; they get together and kick, scream, and wheel and deal, but out of this mess comes the best government in the history of the world. Granted, the quality of government continues to decline, but the fact that we have had 200 years of freedom with this chaotic system is a perfect illustration of non-linearity. Putting bad men in Congress does not necessarily yield a bad government. Likewise, putting "good men" in the Kremlin won't necessarily yield a stable, benevolent republic. The *system* must be dealt with as a whole, for better or worse. Its components are of secondary importance (important in their own right, but definitely subordinate to the whole) when it comes to affecting the quality of government. So it is with biology. The whole system must be dealt with – causes, as well as symptoms.

In their promotional literature, St. Francis Medical Center of California makes the statement that "early detection before symptoms, before a positive chest X-ray, makes the difference in lung cancer. To improve the rate of survival, medicine must begin to diagnose early disease." That has not proven to be true in other forms of cancer, and there is no reason to believe that it will be true with lung cancer. Early diagnosis is not the answer to cancer in any form. The answer is *prevention*. The best hope for a *cure* is wave energy, such as photoluminescence.

Although the previously mentioned Yale technique of photopheresis is unnecessarily complex and expensive, the approach it uses is non-linear. It is therefore more rational than the specific attack on tumor cells engineered in the new laser photo-dynamic therapy. The modified Knott technique (photo-oxidation), which we and a few veterinarians around the country use, takes the non-linear approach of Edelson, but at much less cost.

Zapping cancer cells with infrared laser beams, as in photo-dynamic therapy, is a nice exercise in science, but why attack one point of light when you can use a system, such as photo-oxidation, that will attack "a thousand points of light" (or a million)? The laser-infrared therapy is too complicated, too expensive, and too linear. It certainly has important applications, such as when a bronchial tube is almost obstructed by a tumor mass. At such critical times this type of direct (linear) approach is necessary to shrink the tumor area quickly so the patient can breathe without obstruction. In fact, large obstructive tumors in any part of the body that are accessible to direct visualization by endoscope would benefit from this initial treatment. However, the ultimate treatment of the cancer should be by the non-linear approach of photo-oxidation.

It should also be noted that photo-dynamic therapy with a laser is not without some side effects. These include nausea, vomiting, metallic taste, eye and skin photo-sensitivity, and liver toxicity.

Scientists today just can't get away from the chemical approach to medicine. Photo-dynamic therapy is, as we described, a specific linear approach toward visible tumor cells and does not address the question of generalized metastases (spread) of the disease. However, Manyak, et al, in a review article on photo-dynamic therapy, state: "Generalized metastases could be approached ... using ... a chemical that produces light injected after photo-sensitization." Why not just inject harmless photons into the blood, as in the photo-oxidative method? It is safe, non-toxic and specific for photo-sensitized cancer cells.

Parenthetically, Manyak and his associates list 116 references from the literature, but again, Emmett Knott, the discoverer of blood phototherapy, is overlooked.

British researchers, under the direction of Dr. Raymond Bonnett, also approach the destruction of cancer with high technology. They inject porphyrin into a muscle, wait for the porphyrin to distribute itself throughout the tissues, and then, using an optical fiber directed into the tumor, send laser light down the fiber directly into the tumor tissue. They report that the tumor very quickly dies. But, again, this only cures *the tumor*, not the patient.

Laser light is preferred by the British researchers because this energy can be sent down an optical fiber, and thus can reach deep into tissues and deep-seated cancers. But any source of light, including ordinary tungsten light (as used in the light bulbs in your home), can be used for photo-therapy if the light source can reach the tumor. Some wavelengths of light are more efficacious than others; red light is considered better than the blue end of the spectrum because it has much more tissue penetration.

The laser method is not suitable for large tumors, because even though the laser light penetrates well, it cannot penetrate deeply enough. But once a large tumor has been removed, photo-therapy can then kill remaining traces of malignant tissue in the area of the surgery.

The complex process of photo-destruction of cancer cells is not entirely understood. Porphyrins inside a cell are believed to absorb the introduced red light and become "excited." This excitement is probably caused by the formation of unstable singlet oxygen molecules which then explode the cancer cell.

Ideally, the best photo-sensitizing chemical for cancer treatment would be one that went only to cancer cells and not healthy cells. This is not the way things are in real life and real science, so one has to accept a compromise, using those chemicals absorbed *most* by cancer cells and *least* by normal cells. Chemicals absorb more quickly into cancer cells because cancer cells have a much higher rate of metabolism than normal cells. Because the photo-sensitizing chemical goes to cells throughout the body, it must be a substance that can be cleared rapidly by those cells. Otherwise, the patient will experience cell death from over-exposure to light following administration of the drug.

If the photo-sensitizer is only soluble in fat, such as the bilirubin we mentioned in newborn babies, then it cannot be eliminated from the body and is inadvisable for treatment. However, if it's *too* water-soluble, then the liver will excrete it too rapidly and the photo-chemical will not be in the body long enough to enhance light irradiation.

As red and "infrared" light penetrate tissues better than ultraviolet, the photo-sensitizer should be most effective in the red portion of the spectrum if a tumor is being attacked directly through a fiber-optic tube.

British researchers under Bonnett use an injectable form of photo-sensitizing drug. Dr. Nishimiya of Japan discovered a photo-sensitizer called PH-1126, which is very effectively distributed throughout the body when given by mouth. Both techniques are equally satisfactory.

Although photo-chemotherapy is not yet in the average cancer clinic, the work is progressing rapidly and it should be available in the near future – unless the Food and Drug Administration prohibits its use for political and economic reasons.

As so often happens in modern medicine, Dr. Edelson and his colleagues at Yale have reached back into history for a drug to use in their very high-tech form of photopheresis. The Egyptians recognized, many thousands of years ago, that a plant called ami majus, a weed that grows on the banks of the Nile, had certain interesting medicinal properties that were elicited by exposure to light. The physicians noted that after this plant was eaten, people became very prone to sunburn. They exploited this property in the treatment of a disorder called vitiligo, a condition in which the skin has a blotchy white appearance due to loss of pigment. The active ingredient in the ami plant is a psoralen. Dr. Edelson and his group have been using this plant-derived psoralen – 8-methoxypsoralen, also called 8-MOP.

Psoralens are found in plants other than the Nile weed. Small quantities can be detected in figs, limes, parsnip roots, and other fruits and vegetables. Their qualities are virtually ideal for photochemotherapy: they are generally inert in the human body until exposed to light; they are absorbed quickly from the gastrointestinal tract; they reach a peak level in the blood within two to four hours; and they are completely excreted within 24 hours. Eight-MOP is extremely expensive, so other psoralen derivatives are being explored in the hope of bringing down the cost.

The first modern studies on 8-MOP were done at the University of Cairo, Egypt, in the 1940s. Dr. Abdel M. el Mofty reinvented the wheel when he demonstrated that plants containing 8-MOP, when ingested, would cause a skin reaction when exposed to sunlight. His results came to the attention of a group of scientists at the University of Michigan School of Medicine, who carried out the first studies of purified 8-MOP. Their work, which was done in the early 1950's, proved that the drug was quite safe and that the ratio of the maximum safe dose to that required for clinical effect was very high.

In the 1970's, several investigators showed that 8-MOP, in combination with ultraviolet light, could be used to treat psoriasis. This has become a very effective treatment for this pesky skin condition. Psoriasis, like cancer, is a condition where cells of the skin divide much more rapidly than normal, resulting in thickening and scaling. The principle of cross-linking DNA undoubtedly applies here as well as in cancer and is the reason for the effectiveness of 8-MOP in the treatment of psoriasis. If it works on psoriasis, basically preventing the reproduction of rapidly growing cells, then it should also work on most (if not all) forms of cancer. Applying these methods of treatment – oxygenation of the blood (H_2O_2), photoactive compounds, and exposure of the blood to ultraviolet light of a specific frequency – we have undertaken the treatment of cancer. The results, thus far, have been encouraging.

HOW LIGHT–ACTIVATED DRUGS KILL CANCER

◀ THE PROBLEM IS THAT CANCER CELLS ARE FOUND NEAR HEALTHY CELLS. HOW DO YOU DESTROY THE BAD WITH-OUT HARMING THE GOOD?

▶ WHEN A LIGHT-ACTIVATED DRUG IS INJECTED INTO THE PATIENT, IT IS ABSORBED BY BOTH THE NORMAL AND THE CANCEROUS CELLS

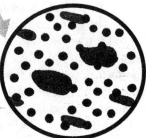

◀ BUT CANCER CELLS MORE READILY RETAIN THE DRUG. AFTER TWO DAYS, IT REMAINS ONLY IN THE TUMOR CELLS

▶ A BURST OF LASER LIGHT THEN MAKES THE DRUG RELEASE A TOXIC FORM OF OXYGEN, DESTROYING ONLY THE CANCER CELLS

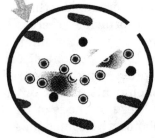

DATA: BW

The Edelson group uses 8-MOP in combination with ultraviolet irradiation of blood in the treatment of cutaneous T-cell lymphoma (CTCL). They found that the cancerous T-lymphocytes are quite sensitive to the effects of 8-MOP in combination with ultraviolet light. With this knowledge, and building on the work of Knott, they treated a series of cases of CTCL with phenomenal results. The average survival of CTCL patients is five years or less. With ultraviolet light treatment, they have achieved very long remissions – ten years or more.

Because neither red blood cells nor platelets have a nucleus, only the white blood cells are affected by the DNA binding effect of the drug. Consequently, Edelson devised a machine which separates the white (nucleated) blood cells from the red blood cells and plasma, and was thus able to concentrate the therapy on the white cells.

His technique irradiates the patient's entire complement of white cells. First, the patient takes the 8-MOP by mouth. After a few hours his blood is withdrawn, the 8-MOP-containing cells are irradiated, and the blood is then reinfused into the bloodstream of the patient.

The white cells are severely injured in this process. The binding effect on the DNA makes the cell abnormal and it subsequently dies. The researchers were concerned about the possible consequences of returning large numbers of damaged and dying T-cells to the patient's bloodstream, but they found that this was actually the key to the success of the therapy. The injured cells stimulate the production of new, healthy cells.

The T-cell is of prime importance in the body's defenses. There are literally millions of different subgroups of T-cells, called clones. These clones are small subdivisions of the T-cell population, each containing a receptor capable of recognizing only a single type of foreign molecule. Any given subgroup is therefore designed to attack only one type of enemy organism. When that enemy presents itself – perhaps a virus or bacterium – the subgroup of T-cells which opposes it rapidly expands as part of the immune response. In a malignancy, however, a clone becomes diseased and expands massively. It eventually dominates the entire white blood cell population, thus killing the patient. The rapid expansion causes the cells to lodge in many areas of the body, including most of the layers of the skin. This invasion is responsible for the characteristic red and swollen skin commonly seen in cases of cutaneous T-cell lymphoma. Because the T-cells are in a malignant state, and are quite active metabolically, they will absorb most of the ultraviolet/8-MOP combination and, when reentered into the bloodstream, cause the body to produce a specific antibody response against that particular line of T-cells. The diseased T-cells often will outnumber the other white cells of the body by a million to one. This causes the spleen to mount a potent immune response specific to T-cells. This remarkable specificity of ultraviolet action to a malignant cell line is a great advance and shows great promise for the treatment of other forms of cancer. Why this immunologic reaction only occurs against damaged cells is not entirely understood.

There are many variations in the techniques originally used by Knott. He used 100 to 300 ml. of the patient's whole blood, whereas Edelson uses 500 ml. at a time and ultimately exposes all of the patient's white blood cells to the ultraviolet light over a three to four hour period. German researchers use a very small amount of whole blood, usually 10 to 15 ml. There is also a very wide range in exposure of these various quantities of blood to ultraviolet light. Knott only exposed the blood for 15 to 20 seconds. The German researchers expose it for four to eight minutes. Yet, with this remarkable range of treatment "doses" (amount of blood multiplied by the time of exposure), they all seem to get quite satisfactory results. Knott and the German researchers noted that quartz crystal must be used in the UV exposure process when using UVC. But Edelson says, "Ultraviolet-A readily passes through glass and certain plastics...." Much research remains to be done. What is the best light source – ultraviolet, infrared, visible light? How much blood should be irradiated? How long should the blood be exposed? What is the most effective photo-sensitive agent?

Infrared light, which has twice the wavelength of ultraviolet, penetrates the skin much more efficiently than ultraviolet, and so should be effective in the treatment of cutaneous T-cell lymphoma by skin irradiation, which is simpler than blood irradiation. The patient could take 8-MOP, and then the entire body could be irradiated with infrared light. It would be interesting to know if the British, who are using this external radiation technique, have tried it on chronic T-cell lymphoma.

Although these methods are significant and will have a great bearing on the treatment of cancer, photo-chemotherapy is not new. It was used for years for the prevention of rickets (vitamin D deficiency) by applying ultraviolet light directly to the skin. Treatment of jaundice in newborn babies, especially premature ones, is highly successful using blue light exposure to the skin of the child.

Although the photosensitive compounds are extremely important and can certainly add to the effectiveness of treatment with ultraviolet light, it is entirely possible that with the use of oxygen the photoactive compounds will not be necessary. After all, Knott's clinical work was done without photoactive compounds (or oxygen, for that matter), yet he had great success in treating infectious diseases. When it comes to killing cancer cells, however, photoactive drugs and oxygen increase the killing effect, and both may be necessary for more effective cancer therapy.

Activated oxygen may have the same effect as the activated photosensitive compounds. The excited singlet state oxygen molecule (O_2) is much more reactive than ordinary oxygen and is known to react with membrane components such as unsaturated lipids, cholesterol and protein. In photosensitized tissues, such reactions would be expected to lead to membrane damage and eventually to cell death. My colleagues and I feel that our success in treating cancer has been due to this activation of singlet oxygen, which is formed in the blood with administration of intravenous hydrogen peroxide ($H_2O_2 \rightarrow H_2O + O_2$). *Activated oxygen may be the main pathway leading to killing photo-sensitized tumor cells.*

In most research centers today, red cells are separated from the blood, leaving only the white cells exposed to ultraviolet light. This is, in my opinion, an unnecessarily expensive variation of the treatment of Knott, which I have found to be extremely effective without any separation of cells. The photo-activation may work on the red cells, or even the plasma proteins, as well as the white cells. So the expensive process of separating the blood into fractions is unnecessary and may actually decrease the effectiveness of the therapy.

Dr. J. L. Matthews, Director of the Baylor Research Foundation, has done research proving that whole blood does not interfere with the penetration of light to the target white cells. Matthews reports:

> "I've tested the herpes virus in full hematocrit whole blood [all red cells present in the sample], to ask the question: Did the presence of the blood interfere either with the light getting to it, or did the presence of whole blood take up so much of the dye itself that it would attenuate, or require a higher dose to kill the virus in terms of dye concentration?

We got comparable viral kill with herpes in the presence of whole blood, so our feeling is that the blood *per se* does not interfere with the system.[3]

It has been proven that iron-containing porphyrin (heme) does not show appreciable fluorescence, nor does ultraviolet light photosensitize singlet oxygen. However, something happens when whole blood is exposed to ultraviolet light. Again, as mentioned, Knott and his associates used nothing but whole blood, and they had quite remarkable clinical results on every conceivable type of infection and toxin. Perhaps the answer is that all of the body's hemoglobin is not tightly bound with iron, and the iron molecule gets "punched out" of the porphyrin ring from the action of ultraviolet light, thus reverting back to a "HpD-type" compound, which is photo-sensitive.

Edelson asked the question: "[H]ow ... can the immune system respond to a group of its own T-cells, malignant or not? The answer takes us out into the deep waters of modern immunology...." The receptor on a T-cell which enables it to recognize specific foreign agents that it is designed to attack is itself a protein that can be recognized by the body's immune system. Through the process of photoluminescence, the T-cell has been turned into a foreign invader from the standpoint of the body's immune system. Edelson reported:

> Since the immune system as a whole is not damaged by photopheresis [Edelson's word for what we call photoluminescence], the most attractive possibility to explain our results is that the immune system (in particular the spleen) mounts a powerful response ... recognizing the malignant clone [of T-cells].

Researcher Irun R. Cohen has found that, by using the photoluminescent method, rats given auto-immune encephalomyelitis (AIE) could transform their T-cells into a "vaccine." When the ultraviolet-treated cells or "vaccine cells" were injected into other rats, those rats could be protected against developing AIE. Thus the treated cells acted as a vaccine against auto-immunity. Edelson concluded from the work of Cohen that photoluminescence must be "vaccinating" T-cell leukemia patients against their own cancer.

---

3 *EIR*, January 29, 1988.

Edelson observed that T-cell leukemia is conquered through this method because the intact immune system stimulates a strong immune response against the treated cells. He said:

> Now it is possible to understand why photopheresis is so effective: the damaged cells of the malignant clone, in effect, prime the immune system to destroy that clone specifically, ridding the body of its cancer.

With this knowledge, it would seem paradoxical that photoluminescence is quite effective in the treatment of AIDS. In AIDS there is a *deficiency* of T-cells, rather than an excess as in leukemia, discussed above. The T-cell count will often go down to almost zero in AIDS patients. This being true, it doesn't seem likely that activation of abnormal T-cells could play a part in the treatment. The answer to the effectiveness of the treatment may lie in the fact that we are also using singlet oxygen by way of intravenous peroxide. The singlet oxygen yields positive results by killing off all circulating virus and by other mechanisms not yet understood. Perhaps the therapy will not work in AIDS patients who do not have enough T-cells left to form an "immune pool" of cells to which the body can respond. As of now, we simply don't know.

Summary of Clinical Reports of Photodynamic Therapy

Type	No. of Patients	Response Complete Recovery*	Partial Recovery**	No Response***	Follow-up (months)	Side Effects	Comments
Lung, advanced	300	250	30	20	to 24	hemoptysis, fistula, resp. distress	Complete recovery= opening of bronchus
Lung, early	70	28	31	11	to 70		28 patients were x-ray occult
Bladder cancer	57	50	3	1	to 32 (none in 3)	irrative symptoms, frequency, urgency, spasm	most by whole bladder treatment
Esophagus, advanced	17	0	11	6	to 12	mediastinal stenosis	Partial recovery = pain, fistula, partial opening of obstruction
Head/neck, advanced	147	53	56	37	to 24	pain, hemoptysis, edema	most treated after surgery and/or x-ray therapy
Head/neck early	100	45	30	25	to 19	—	unpredictable response
Skin (metastatic) breast, basal cell, squamous cell, melanoma	(219)	(147)	(20)	(43)	to 48	pain, ulcer	most treated after surgery, x-ray therapy, chemotherapy
Gyn****	41 (86)	(49)	(25)	(6)	to 28	pain, discharge	most were recurrent
Ocular (melanoma, retinoblastoma, periocular)	(60)	55	most cases; floating retinoblastoma, unresponsive		to 24	Tumor shrinkage in acuity, retinal detachment, cataract hemoptysis	loss of visual side effects in most patients resolved in days to weeks.

*no visual tumor
**decreased at least 50% of original volume
***less than 50% reduction in tumor volume
****Not explained by contributor—probably cancer of the uterus and/or cervix.

Reference: McCaughan and Dougherty, Postgrad. Gen. Surg., September, 1989

Chapter 6

Photoluminescence in Action: Case Studies by the Pioneers

The worth of any therapy is truly in its results on patients, not on the blackboards at med school or the pages of books. So one of the best things to do with regard to photoluminescence is to look at its record. It has been used so often, with such uniformly beneficial results, that its record on patients is the best possible testimonial to its efficacy. Thanks to the pioneers of this therapy, thousands of successful cases are available for study. Following are a few of them.

A Summary of the Cases of Henry A. Barrett, M.D.

Dr. Barrett reported on 110 cases of ultraviolet light therapy in 1940.[1] Of those 110 patients, some received only one treatment and others as many as eight. The report indicated that outstanding results were often obtained with a single irradiation, but sometimes no improvement was noted until more had been received. Most of Dr. Barrett's cases had not responded to any other type of treatment. He felt that his statistics would have been better had several patients not become discouraged after not improving on the first treatment, and so discontinued the therapy. Barrett remarked that in conventional therapy one does not expect to observe any appreciable change in a patient's condition in under twelve hours, and often not until several days after initiation of therapy. But with hemo-irradiation,

---

1 Medical Clinics of North America, May 1940.

he noted patients suffering from rheumatoid arthritis improved, remarkably, within a few hours. One of his patients came to him for a condition completely unrelated to her arthritis, but after three treatments, said her arthritis had cleared completely. She had come for treatment of her night sweats, which had necessitated a change of linen three and four times a night, winter and summer. This condition also cleared after the third hemo-irradiation.

An illustration of the remarkably wide efficacy of ultraviolet treatment involved a patient referred for chronic inflammation of the eyelid. The patient had been treated by an eye specialist for two years with absolutely no positive result. Barrett initiated ultraviolet irradiation therapy and the inflammation cleared in less than a week.

Dr. Barrett describes another quite remarkable case:

> In March of this year, one of my associates in this investigation assisted me in the irradiation of a patient of his suffering from bronchial asthma of four years duration. The patient, aged 45, had been in the hospital for several weeks and, in spite of all medication, was having several attacks daily. During the irradiation she had two attacks. Her doctor reported the next day that she had had only one attack in 24 hours. After that, she had an occasional relapse, but not more than two or three a week. The attacks became fewer and fewer and have been entirely absent for several months.

Toxemias, Dr. Barrett remarked, almost always clear up quickly following hemo-irradiation. Dr. Barrett summarized his 110 cases as follows:

1. No detrimental reactions from ultraviolet irradiation have been observed. With the factors employed, it is a very safe method.
2. Improvement is frequently almost immediate.

"Physiologic effects," he commented, "commonly may be summarized as follows:

1. Increase in peripheral circulation.
2. Increase in the combining power of blood in oxygen.
3. Inactivation of toxin in the blood."

Some of Dr. Barrett's results were truly astounding. A patient with mastoiditis had been ill for 19 days. X-rays showed "a four-plus mastoid infection with coffining." She refused an operation but consented to hemo-irradiation. Without surgical intervention, this degree of mastoid infection would ordinarily lead to meningitis, brain abscess, and death. But she was back at work in six days, entirely symptom-free, and remained so thereafter.

Dr. Cecil Rountree remarked many years ago on the effectiveness of ultraviolet treatment ("sunlight treatment, real or artificial"). He noted that extra-pulmonary tuberculosis (beyond the lungs), such as bone tuberculosis, cleared very nicely, without drastic surgery, with complete rest and a great deal of sunlight. This method "almost always effects a cure in every case," he said.

Tuberculosis is still a common scourge in most tropical countries in spite of an abundance of sunlight (black skin absorbs light poorly). Ultraviolet blood irradiation, far more effective than simply irradiating the skin as Rountree and Finsen did, will be a great boon to Africa, Asia, and Latin America in treating this debilitating and often fatal disease.[2]

Rountree remarked:

> The skin has certain limitations as far as ultraviolet irradiation treatment is concerned. The thickness, natural color, pigmentation, and other factors lead to a great variability in the amount of absorption that can be obtained from ultraviolet skin irradiation. In addition to these factors, there are inherent faults of the skin as a receptor for ultraviolet irradiation as, for example, the person's sensitivity and susceptibility to radiation injury. Notwithstanding these natural handicaps, which limit efforts to utilize the skin as a medium for the absorption of ultraviolet energy, much has been accomplished in using ultraviolet irradiation of the skin for a number of conditions.[3]

The Cases of Dr. George P. Miley

Researching in the 1930s and '40s, Dr. Miley reported some interesting cases which he said resulted in "meager but stimulating observations." His observations were not meager at all, but were certainly stimulating.

For instance, in 11 patients suffering from severe and intractable furunculosis (boils) of up to six years duration, he

---

2 Since this was written, TB has become recognized as a serious public health problem in the U.S.

3 Rountree, *Encyclopedia Britannica*, 14th Edition, Vol. 3, p. 846.

noted a complete subsidence of the acute stages of the disease and a lack of recurrences after two to four blood irradiation treatments given at two- to five-week intervals. One of these patients had suffered from the daily appearance of 40 to 60 new boils for years. This patient, after only four blood irradiations, had a complete subsidence of his disease. He was placed on a maintenance dose of two to three blood irradiations a year.

Miley also reported on six patients with intractable herpes zoster (shingles). This is a virus similar to the chicken pox virus which causes painful blisters on the skin, usually at the belt-line in men and at the bra-line in women. Conventional treatment is generally ineffective.

Extracorporeal photo-therapy often succeeds in saving the patient's life, where antibiotic therapy fails. A summary of work done in the 1940s, using the therapy in hopeless cases, proved that blood ultraviolet irradiation controls acute bacterial infections, with or without septicemia (blood poisoning) to an extent hitherto never dreamed imaginable. While Staphylococcus (a virulent bacterium when in the blood) and endocarditis (an infection of the valves of the heart) did not respond well, it was almost always due to the inhibiting effect of sulfa chemotherapy.

In January, 1942, Dr. George Miley made the following observations:[4]

> The detoxification effect of ultraviolet is not generally known by the medical profession and certainly has not been emphasized enough. The inactivation of snake venoms and bacterial toxins are examples of what may be accomplished by ultraviolet.
>
> The increased ability of blood irradiated with ultraviolet to absorb oxygen has been demonstrated.[5]
>
> As a rule, rather low dosages of externally applied ultraviolet radiations stimulate the general resistance of animals and human beings to infection.[6]

---

4 *N.Y. State Journal of Medicine*, January 1, 1942.

5 Mayerson, H.S. and Laurens, H.: *Journal of Nutrition* 3: 465 (1931); Miley, G.: *American Journal of Medical Science* 197: 873 (1939).

6 Clark, J.H., Hill, C., Handy M., Chapman, J., and Donahue, D.D.: *Ultraviolet Radiation and Resistance to Infection*, Copenhagen, Congres. Internat. Lum., 1932, pp. 458-463.

Table—Results in Cases of Acute Pyogenic Infection Given Ultraviolet Blood Irradiation Therapy at the Hahnemann Hospital, Philadelphia, from November 1, 1938 to April 1, 1941*

	No. of Cases	Recovered	Died
Early:			
Puerperal sepsis	2	2	
Incomplete septic abortion	2	2	
Acute ulcerative gingivitis secondary to third molar abscess	2	2	
Acute furunculosis or carbunculosis	7	7	
Acute Streptococcus hemolyticus oropharyngitis	1	1	
Acute pansinusitis	1	1	
Acute tracheobronchitis	1	1	
Acute pyelitis	1	1	
Wound infections	2	2	
Fever of undetermined origin	1	1	
Moderately Advanced:			
Puerperal sepsis	7	7	
Incomplete septic abortion	8	8	
Pelvic abscesses; pelvic peritonitis	7	7	
Peritonitis, generalized	10	10	
Wound infections	3	3	
Acute femoral thrombophlebitis	3	3	
Acute Streptococcus hemolyticus oropharyngitis	1	1	
Fever of undetermined origin	2	2	
Bronchopneumonia	1	1	
Acute osteomyelitis, advanced nephrosis	1		1
Acute cholecystitis, cholelithiasis	1	1	
Double otitis media	1	1	
Streptococcus viridans septicemia secondary to parotitis	1	1	
Acute suppurative hemorrhagic cystitis	1	1	
Apparently Moribund:			
Puerperal sepsis	2	2	
Incomplete septic abortion, hemorrhagic shock	2		2
Peritonitis, generalized	1	1	
Appendiceal abscess	1	1	
Pelvic abscesses, pelvic peritonitis	5	4	1
Wound infections	1	1	

* *The New York State Journal of Medicine,* January 1, 1942.

Continued from page 73

	No. of Cases	Recovered	Died
Fever of undetermined origin	2	1	1
Lobar pneumonia	2	1	1
Bronchopneumonia	1	1	
Pyelonephritis, cystitis, secondary to bladder carcinoma	1		1
Rectal abscesses, cystitis, ileitis, advanced arteriosclerosis	1		1
Bacillus coli abscess of scrotum	1	1	
Streptococcus hemolyticus oropharyngitis complicating mastoidectomy	1	1	
Extensive trauma, terminal bronchopneumonia	1		1
Septicemias:			
Staphylococcus aureus	4		4
Staphlococcus albus secondary to Staphylococcus albus pneumonia	1		1
Streptococcus hemolyticus	2	2	
Streptococcus nonhemolyticus	2	1	1
Streptococcus viridans subacute bacterial endocarditis	4		4
Streptococcus nonhemolyticus endocarditis	1		1

Summary:

	Early	Moderately Advanced	Apparently Moribund
Number of cases	20	47	36
Number recovered	20	46	17
Percentage recovered	100	98	47
Number died	0	1	19
Percentage died	0	2	53

All of the patients in the *apparently moribund* group were probably on sulfa therapy, which renders photoluminescence ineffective.

While artificially-induced infections in laboratory animals can be helpful for determining the efficacy of a treatment, there is no substitute for seeing the effect in human beings. The following cases illustrate the dramatic drop in fever in severely ill patients and are a graphic illustration of the remarkable efficacy of this treatment.

Miley presented a rare case of "double septicemia" in which blood cultures were found to be positive for both streptococcus and a colon bacillus following a prostate operation. This patient was obviously in deep trouble and was going to die. Antibiotics had not been effective.

Three blood photoluminescent treatments were given at 48-hour intervals. Forty-eight hours after the first treatment, the blood cultures became negative, i.e., the bacteria could no longer be recovered from the blood. This proved to be only temporary. The streptococcus infection disappeared permanently, but the colon bacteria reappeared in the blood just before a third blood irradiation was given. The patient's temperature rose to 106.8°F. Following the third irradiation, the patient's condition improved greatly and blood irradiation treatments were continued.

It was interesting to note that blood cultures continued to be positive with the colon bacillus for three days following the last treatment, even though the patient continued to improve. The patient progressed without any further complications and was discharged from the hospital feeling perfectly well, with all blood tests and cultures within normal limits.

The second case was a blood infection from a colon bacillus secondary to an acute infection of the urinary bladder. When the patient was first seen, she had been in a coma for 40 hours. Her blood cultures had been positive for 13 days with a temperature that had reached the astronomical level of 108.4°F.

Forty-eight hours after the initial hemo-irradiation, the blood cultures became negative and the patient's toxic symptoms rapidly subsided. Ninety-eight hours after starting treatment, the temperature was back to normal and the patient was feeling perfectly well. This patient surely would have died without photoluminescent therapy.

The third case was streptococcus blood poisoning (septicemia), again with an astronomically high temperature. The septicemia was secondary to an ear infection (otitis media). When first seen, the patient had been in a coma for two days, and the temperature had reached 108.2°F. Forty-eight hours after the initial blood irradiation, the blood culture became negative, but 72 hours later, the blood culture was positive again. The patient began to deteriorate, so a second blood irradiation was given. Within 48 hours, blood cultures were again negative. The patient recovered with no further problems.

The fourth case was blood poisoning following a Caesarean section operation. At the time the patient was first seen, she had undergone 22 days of intensive antibiotic therapy, which had not stopped the infection. Twenty-four hours after her first treatment, the blood cultures became negative, and 48 hours later, the patient was free of all symptoms for the first time in 26 days.

The last case history is one of an incomplete septic abortion (a miscarriage with only partial expulsion of the foetus and the placenta) with infection in all the female organs. The patient was moribund when first seen and had a temperature of 106.4°F. Despite eight days of intensive antibiotic therapy she had gotten worse. On the first day after her ultraviolet irradiation, her temperature dropped precipitously and her toxic symptoms began to subside. Forty-eight hours later, her detoxification was complete. The patient had completely recovered, without any further treatment, nine days after beginning her blood irradiation.[7]

Six months after the above report, Doctors Miley and Rebbeck reported a large series of cases in the *Review of Gastroenterology*.[8] The sensational cures are of such monotonous regularity that we have only reported a few typical cases from their series. The tables on pages 78-86 summarize their amazing results.

Clinical Observations by Dr. Miley's Peers at Hahnemann Hospital

We wish to state unequivocally that in the more than 6,000 ultraviolet blood irradiations administered during the past four

---

7 These cases were from Shadyside Hospital, Pittsburgh, PA; Portland Sanitarium, Portland, OR; Bellevue Hospital, New York; Hahnemann Hospital, Philadelphia; and City Hospital, New York, as reported by Miley in *Archives of Physical Therapy*, September 1942.

8 *Review of Gastroenterology*, January-February, 1943.

years at the Shadyside Hospital in Pittsburgh and at the Hahnemann Hospital in Philadelphia we have seen no deleterious effects following the use of this method, either from a clinical or a laboratory standpoint.

In many of the cases reported here in which recovery occurred there was observed at the time of initial ultraviolet blood irradiation therapy a definite paralytic ileus (intestinal paralysis causing a loss of function), as might be expected. The ileus had been present as long as four days in some instances, but in all cases of recovery reported, the ileus disappeared in very rapid dramatic fashion after ultraviolet blood irradiation therapy at intervals varying from 12 hours to 96 hours. The disappearance of the paralytic ileus and the reappearance of normal intestinal smooth muscle tone was marked by the expulsion of large amounts of flatus, a marked reduction in abdominal distention, and normal auscultation [diagnostic monitoring of the sounds of an internal organ] findings of the abdomen, accompanied by a subsidence of other toxic symptoms. This dramatic abolition of paralytic or adynamic ileus has constituted one of the chief clinical observations made during the course of our work with blood irradiation in peritonitis.

It has been our impression that the optimum time for the application of ultraviolet blood irradiation therapy in all cases of peritonitis is before operation. Certainly the sooner it is instituted, the better.[9]

Dr. H. M. Eberhard remarked:

At the Hahnemann Medical College and Hospital in Philadelphia, the Committee on Research is very exacting. Before any research report is released for publication it must be approved by the Committee.

In the investigation submitted by Dr. Miley today the Committee postponed release a long time, until assured that the premises and conclusions warranted their stamp of approval.

No doubt many of you have wondered what advantage, if any, blood irradiation had over sulfanilamide and its derivatives, and since the sulpha products appear to have been

---

9 Miley, G.: *American Journal of Physiology*, 388, June 1941.

Miley reported the following summary of cases in 1947:*

Results in 445 Cases of Acute Pyogenic Infection Given Ultraviolet Blood Irradiation Therapy At the Hahnemann Hospital, Philadelphia, Over a Period of Six and One-Half Years**			
	No. of cases	Recovered	Died
Early			
Puerperal sepsis	3	3	
Incomplete septic abortion	12	12	
Acute furunculosis and carbunculosis	15	15	
Abscesses	2	2	
Acute Strep. hemolyticus	5	5	
Acute tracheobronchitis	12	12	
Acute wound infections	2	2	
Acute pansinusitis	1	1	
Acute pyelitis	1	1	
Acute otitis media	2	2	
Fever of unknown origin	1	1	
Moderately advanced			
Puerperal sepsis	14	14	
Incomplete septic abortion	57	57	
Acute furunculosis and carbunculosis	21	21	
Apparently moribund			
Puerperal sepsis	4	3	1
Incomplete septic abortion	2	1	1
Generalized peritonitis	4	2	2
Abscesses: Pelvic	6	5	1
Rectal	2	1	1
Scrotum	1	1	
Wound infections	3	2	1
Fever of unknown origin	2	1	1
Lobar and bronchopneumonia	2	1	1
Tb. meningitis	3		3
Mesenteric thrombosis, diabetes mellitus	1		1
Septicemias:			
Staph. aureus—albus with sulfa drugs	7		7
without sulfa drugs	9	9	
Str. hemolyticus	3	3	
Str. non-hemolyticus	2	1	1
Str. viridans—subacute bacterial endocarditis	13		13
Str. hemolyticus endocarditis	1		1
Str. non-hemolyticus endocarditis	1		1

* *Am. J. Surgery*, April, 1947.

Continued from page 78

	No. of cases	Recovered	Died
Miscellaneous Cases:			
Abscesses	13	13	
Acute Strep. hemolytic pharyngitis	5	5	
Endometritis and parametritis	17	17	
Acute wound infections	6	6	
Salpingitis	15	15	
Peritonitis	18	16	2
Osteomyelitis	16	15	1
Fever of unknown origin	13	13	
Lobar and bronchopneumonia	11	11	
Atypical (virus) pneumonia	10	10	
Preoperative	13	13	
Postoperative	44	41	3
Non-healing wounds	6	6	
Thrombophlebitis	34	34	

Summary*

	Early	Moderately Advanced	Apparently Moribund
Number of cases	56	323	66
Number recovered	56	317	30
Percentage recovered	100%	98%	45%

Most of the patients in the *apparently moribund* group were probably on sulfa therapy which renders photoluminescent therapy ineffective.

*Not all cases treated are listed -Editor

Report on Detoxification Effect Following Ultraviolet Blood Irradiation in Generalized Peritonitis**

Case No.	History No.	Age	Sex	Condition* at time of initial blood irradiation	Detoxification time** Early	Complete	No. of irradiations	Result***
1	H 55321	44	F	M.A.	24 hrs	72 hrs	1	R
2	S 90679	55	M	M.A.	24	48	1	R
3	H 67575	20	M	A.M.	24	48	1	R
4	S 90923	17	M	M.A.	24	72	2	R
5	H 55762	34	M	M.A.	48	72	1	R
6	S 91913	35	F	M.A.	24	48	1	R
7	H 37339	52	M	A.M.	24	120	3	R
8	S 74019	11	F	M.A.	48	72	1	R
9	H 55738	11	F	M.A.	24	48	1	R
10	S 71033	17	M	A.M.	24	48	1	R
11	H 59266	17	M	M.A.	48	120	2	R
12	S 71997	21	M	A.M.	24	72	1	R
13	H 57010	59	M	M.A.	72	96	1	R
14	S 75699	24	F	M.A.	48	120	1	R
15	H 69619	22	F	M.A. with strep. hemoly-ticus septicemia	24	240	4	R
16	W.J.H.	28	F	M.A.	24	96	2	R
17	H 53365	26	M	A.M.	No detoxification		1	D
18	H 62542	16	F	M.A. not operated	48	96	2	R
19	H 57356	25	F	A.M.	48	96	1	R
20	S 98831	35	F	A.M.	24	96	2	R
21	H 48302	48	M	M.A.	24	72	1	R
22	S 84504	49	F	M.A.	48	96	3	R
23	S 88704	13	M	M.A.	24	48	1	R
24	S 88985	34	M	M.A.	24	72	1	R

25	S89361	40	F	M.A.	48	72	1	R
26	S89551	14	M	A.M.	48	72	1	R
27	S91684	25	F	A.M.	48	96	2	R
28	S92929	22	M	M.A.	24	48	1	R
29	S94739	47	F	A.M.	24	72	2	R
30	S96073	28	F	M.A., with abcess	48	72	1	R
31	S96140	8	M	M.A.	24	72	1	R
32	S96227	14	M	A.M.	48	96	2	R
33	S78087	57	F	A.M.	No detoxification		3	D
34	S77119	66	M	A.M.	No detoxification		3	D
35	S77948	86	M	A.M.	No detoxification		1	D
36	S76608	19	F	A.M.	No detoxification		1	D
37	S81817	54	M	A.M., atelectasis, cardiac failure	No detoxification		1	D
38	S86406	54	F		No detoxification		2	D
40	H42036	27	F	M.A.	24	48	1	R

Summary

23 out of 23 Moderately Advanced - 100 % recovered

9 out of 17 Apparently Moribund - 53% recovered

26 received only one irradiation
9 received two irradiations
5 received more than two irradiations

Complete detoxification in 32 cases

No detoxification in 8 cases

Average detoxification time:
Early   34.5 hours
Complete 81.86 hours

*See Key for next table

**Miley and Rebbeck, *Rev. Gastroent.*, Jan-Feb, 1943.

Report of Detoxification Effect Following Ultraviolet Blood Irradiation*
in Localized Peritonitis, Appendiceal Abscess

Case No.	History No.	Age	Sex	Condition* at time of initial blood irradiation	Detoxification time** Early	Complete	No. of irradiations	Result***
1	S 83105	32	M	A.M.	24	48	1	R
2	S 84113	17	M	M.A.	48	72	1	R
3	S 72507	41	M	A.M.	24	48	1	R
4	S 71301	22	M	M.A.	12	48	1	R
5	S 71542	11	M	M.A.	24	48	1	R
6	S 72650	78	F	M.A.	24	48	1	R
7	S 76708	14	F	M.A.	24	72	1	R
8	S 84919	22	M	M.A.	24	48	1	R
9	S 85358	19	F	M.A.	24	48	1	R
10	S 85470	49	M	M.A.	24	72	1	R
11	S 85508	40	M	M.A.	24	48	1	R
12	S 85513	47	M	M.A.	24	36	1	R
13	S 85520	51	M	M.A.	24	48	1	R
14	S 87859	38	M	M.A.	24	48	1	R
15	S 91608	35	F	M.A.	24	48	1	R
16	S 87689	67	F	A.M. (uremia)	No detoxification		2	D
17	S 87879	52	F	A.M.	No detoxification		5	D
18	S 89539	38	M	A.M. (delirium tremens)	No detoxification		2	D
19	H 59398	26	M	A.M.	24	72	2	R
20	H 63468	30	F	A.M.	24	48	2	R

Summary

Complete detoxification in 17 cases
No detoxification in 3 cases
Average detoxification time:
  Early   24.71 hours
  Complete 52.94 hours

13 out of 13 Moderately Advanced - 100% recovered
4 out of 7 Apparently Moribund - 57% recovered
15 received only one irradiation
4 received two irradiations
1 received more than two irradiations

*M.A. — Moderately Advanced, in which temperature, pulse and respiratory rates exceed 101-102 F., 100-110 and 24-25 respectively, and such toxic symptoms as nausea, vomiting, restlessness, irritability and mental confusion are excessive.

*A.M - Apparently Moribund, in which the symptoms present are a combination of those advanced symptoms commonly considered near terminal or terminal, namely, coma, rapidly falling blood pressure in some instances, cardiac irregularity, irregular and shallow respiration, obvious loss of thermotactic control, and often an associated septicemia.

**The detoxification time is arbitrarily divided into two components, early and complete, which are defined respectively as: first, the time in hours which elapses between initial blood irradiation and early signs of subsidence of toxic symptoms; second, the time in hours elapsing between initial blood irradiation and a complete detoxification effect, i.e., that point at which it becomes clearly evident that the patient's condition is no longer precarious.

*** R— Recovered; D— Died.

Report of Detoxification Effect Following Ultraviolet Blood Irradiation in Pelvic Peritonitis, Multiple Pelvic Abscesses

Case No.	History No.	Age	Sex	Condition* at time of initial blood irradiation	Detoxification time**		No. of irradiations	Result***
					Early	Complete		
1	S 88538	33	F	A.M.	48	72	1	R
2	H 40602	43	F	M.A.	120	144	2	R
3	H 52080	26	F	M.A.	24	72	1	R
4	H 52365	28	F	M.A.	24	48	1	R
5	H 55364	25	F	A.M.	48	96	2	D (operative shock)
6	H 36760	45	F	A.M.	24	36	1	R
7	H 37960	23	F	A.M.	96	120	2	R
8	H 51921	33	F	A.M.	48	96	2	R
9	H 54003	40	F	A.M.	48	96	1	R
10	H 67446	27	F	A.M.	96	120	2	R
11	H 56102	23	F	A.M.	24	None	3	D (sigmoid carcinoma)
12	H 64296	34	F	A.M.	48	None	2	D (sigmoid carcinoma)

Summary

Complete detoxification in 10 cases	3 out of 3 Moderately Advanced - 100% recovered
Incomplete detoxification in 2 cases	6 out of 9 Apparently Moribund - 67% recovered
Average detoxification time:	5 received only one irradiation
Early	54 hours	6 received two irradiations
Complete	90 hours	1 received more than two irradiations

*M.A.– Moderately Advanced, in which temperature, pulse and respiratory rates exceed 101-102°F., 100-110 and 24-25 respectively, and such toxic symptoms as nausea, vomiting, restlessness, irritability and mental confusion are excessive.

*A.M.– Apparently Moribund, in which the symptoms present are a combination of those advanced symptoms commonly considered near terminal or terminal, namely, coma, rapidly falling blood pressure in some instances, cardiac irregularity, irregular and shallow respiration, obvious loss of thermotactic control, and often an associated septicemia.

**The detoxification time is arbitrarily divided into two components, early and complete, which are defined respectively as: first, the time in hours which elapses between initial blood irradiation and early signs of subsidence of toxic symptoms; second, the time in hours elapsing between initial blood irradiation and a complete detoxification effect, i.e., that point at which it becomes clearly evident that the patient's condition is no longer precarious.

accepted generally by physicians and surgeons, that question confronted the Research Council and was discussed in detail.

It was difficult at times to draw a line of demarcation, since each method approached the other in many instances with similar results. One advantage which was very noticeable was the complete absence of reactions when irradiation was used. I have never seen a single reaction in an afebrile [no fever] case. In generalized infection where nausea is an outstanding symptom, the patient did not have an aggravation, as often occurs with the sulfanilamide derivatives. Very good results were noticed in my clinic in cases of inanition [hunger exhaustion]. We all have seen the patient who is extremely thin, with anorexia, anemia, etc., and on whom diets well indicated make very little or no difference in increasing weight. A number of such cases were investigated and found to have a very low venous oxygen. Accepting the normal as 12 to 15 volumes percent, we found the inanition cases had as low as 2 to 5 volumes per cent. After several irradiations six of eight cases gained as much as ten and fifteen pounds, with a corresponding increase in the oxygen-carrying capacity of the red cells.

I have witnessed the dramatic results in cases of puerperal sepsis, lobar and bronchopneumonia, etc.

In one case of proved undulant fever (brucellosis) that had been refractory to the usual forms of therapy, no fever was noted after the first irradiation.

While we well understand that one case is not sufficient to report, the unusual results in the case mentioned were so startling that Dr. Miley hopes to treat as many such cases as possible and make a complete report.

Substantially, the best results noticed by many when using hemo-irradiation were in lobar pneumonia, bronchopneumonia, influenza and generalized infections. It has been helpful in most instances in surgical preparation, and especially in abdominal cases the patient was distinctly less disturbed and the convalescence uneventful.

In some cryptogenic temperatures, where no causative factor could be detected, it was found to be helpful.

Where habitus enteropticus existed, with collateral weakness of the non-striated muscular system, many patients were benefitted to an unusual degree.

In ulcerative colitis it has not given satisfactory results until after the colon was put to rest, preferably by a double-barreled colostomy.

I was much impressed in cases of hypoproteinemia, especially in those in which the albumin-globulin ratio balanced.

My experience with this therapeutic agent is such that I feel that it warrants consideration and, in many instances, may be helpful when other means of therapy have failed.

The Rebbeck Studies

Dr. E. W. Rebbeck was another of the great pioneers of ultraviolet light therapy. After years of practicing this treatment in Pennsylvania, he was able to report:

> If the conception of septicemia is the presence of pathogenic organisms in a patient's blood plus septic symptoms, then seven patients with Escherichia coli (E. coli) septicemia [a fecal contamination of the blood] have been successfully treated in the Shadyside Hospital by the Knott technique of ultraviolet irradiation of blood.
>
> The rationale of blood irradiation therapy is based on accepted biophysical effects of ultraviolet rays, chiefly those of detoxification and inactivation of bacteria, toxins and viruses. From a clinical point of view, any therapeutic measure that raises the patient's resistance to infection ... should be useful as an adjunct in the treatment of infections.

Rebbeck continued:

> The physical principle that any substance which is capable of absorbing ultraviolet rays (such as blood is known to be) gives off secondary emanations, would explain the ultimate destruction of bacteria in the blood. The commonly-seen detoxifying action of ultraviolet rays, as reported in the references cited, explains the beneficial effect of this therapy in helping to overcome infection.

This Rebbeck report of 1943 concerned the treatment of E. coli septicemia. He reported on eight cases, out of which six lived – a success rate far beyond what could have been expected at that time from the use of sulfa drugs.

It is interesting to note that of the two patients who died, one was reported at autopsy to have a sterile bloodstream, and the other showed only one infection in the bloodstream instead of the two cultured before death. For reasons unknown at the time, the remaining infection, staph aureus, did not respond as well to photoluminescent therapy as other bacteria.[10]

The amazing thermolytic (temperature-lowering) effect of ultraviolet blood irradiation on the body is well-illustrated by Rebbeck's studies on E. coli septicemia. Let's look briefly at three of his more sensational cases:

"Mr. B" was a white man, age 76, who was admitted to Shadyside Hospital complaining of pain on urination, frequency of urination and, occasionally, an inability to urinate. Laboratory studies revealed that he had a 19,750 white count (normal is 5,000 to 10,000). A suprapubic cystotomy[11] was performed and his condition remained stable for a few days; a suprapubic prostatectomy[12] was then performed. He did well after those surgeries, but then developed a chill and his temperature rose rapidly to 105.8°F.

At that time, he was given his first photoluminescent treatment. A dramatic, almost unbelievable, immediate drop in his temperature was noted. It fell from 105.8 to 98.6. In the next few days, this dramatic response to ultraviolet blood irradiation was repeated a number of times. After the third dramatic thermolytic event, his temperature gradually returned to normal and stayed within normal limits, except for occasional small increases.

With a urinary catheter change on day 40, he again had a high temperature for two consecutive days. Following another hemo-irradiation treatment, his temperature rapidly plunged back to normal. After ten days he was discharged, having recovered completely from this serious illness *without the use of any antibiotics*.

An equally dramatic case of bilateral pyelonephritis[13] was reported by Rebbeck. The infection, E. coli, had entered the bloodstream.

The patient was given a total of 15 blood irradiations. He was also given sulfa drugs (discontinued because of side effects), sodium iodide and intravenous aspirin.

---

10 A year after this study, Miley solved "the mystery of the photon-resistant Staphylococcus" in a report in the *American Journal of Surgery*.

11 A drain tube from the bladder placed just above the pubic bone.

12 Surgical removal of the prostate.

13 Severe, and often fatal, infection of the kidneys.

The patient had an extremely stormy course with days of delirium, marked toxemia, and intense pain and tenderness over both kidneys. After the first irradiation, the kidney pain practically disappeared. Although the patient was considered hopeless by his attending physicians, after 25 days he was discharged completely well. As could be seen from his temperature charts, phototherapy never failed to improve his condition, and his life was undoubtedly spared by Rebbeck's treatment.

Rebbeck's most dramatic case was one of pyelonephritis in which the temperature "went through the roof."

After a cystoscopic examination, the patient's temperature quickly rose to 109°F. The patient's pulse rate was 140; she had pain in her joints and felt extremely weak, lethargic and apprehensive. She began to vomit blood, and it was felt that she almost certainly would die.

Blood irradiation therapy was instituted, and, again, as in his previous cases, a dramatic fall in temperature and marked clinical improvement was noted. After a few more days she was brighter, her chills had ceased, all toxic symptoms were gone, and she was well on her way to recovery. This woman was saved literally by a few pennies' worth of electricity.

Rebbeck commented on these cases:

> There have been no signs of harmful effects in approximately 4,000 blood irradiation treatments under my direct supervision at Shadyside Hospital in the past five years. If one looks at the over-all picture of serious infections and realizes that accepted methods of therapy fall far short of producing consistently good results, one appreciates that there is room for any logical, harmless type of therapy such as the ultraviolet irradiation of the blood of patients.

After presenting these three astounding cases, a number of physicians at the conference made interesting remarks concerning the work of Rebbeck.

Dr. G. J. P. Barger (Washington, D.C.) said:

> He has been able to save five out of seven cases of Escherichia coli septicemia, which by all previous experience, would have all been fatal. The results in septic abortions have been so impressive that at one large hospital where this service is available, the ultraviolet irradiation of auto-transfused blood has been made standard treatment for all abortion cases. It is effective in all the conditions for which the sulfa drugs are used; it is successful in many of

the conditions after the sulfa drugs have failed; it is not followed by any of the damaging effects that are common to the sulfa drugs; it is effective in a wider range of bacterial and virus infections than are the sulfa drugs; it is effective in inactivating various types of destructive toxins; and it produces a wide-spread and prolonged dilatation of peripheral capillaries.

Dr. Roswell Lowry (Cleveland, Ohio):

Dr. Rebbeck has shown that ultraviolet blood irradiation therapy is effective after the sulfonamides have failed. This has been my experience. We have given more than 200 irradiations to 100 patients. These patients have been given a thorough course of sulfonamides with no success. I have seen angry, edematous infections subside in six to eight hours following blood irradiation therapy and in 24 hours they'd be almost gone.

Dr. Disraleai Kobak (Chicago, Illinois):

Perhaps the most significant fact to appreciate is that a concrete and objective observation has been repeatedly recorded concerning a new procedure of applying ultraviolet energy, so that it challenges the very latest chemotherapeutic products to ameliorate or even cure those moribund states in which these have failed.

In his closing remarks, Dr. Rebbeck discussed chylomicrons, the tiny fat particles in the blood which are so important for general health in resisting infection. Rebbeck said:

In many instances we were able to convert a greatly disturbed micron picture to normal. For instance, a picture of 15-20 chylomicrons clumped together with great variation in size and shape and no Brownian movement (the dancing motion of minute particles suspended in a liquid), often changed to normalcy after five to fifteen seconds exposure to ultraviolet. The normal picture would be that of no clumps, good Brownian movement and all kinds of chylomicrons approximately the same size (about 1/2 micron) and shape.

If, as has been claimed, the function of chylomicrons is absorbing toxins, this action might well be very significant and be a factor in explaining the critical detoxifying action of blood irradiation therapy.

Dr. Rebbeck then commented on why he continued to endure the slings and arrows of outrageous medical bigotry to continue phototherapy:

The reason I persisted and continued work in blood irradiation therapy for five years, in spite of much opposition and criticism from the medical profession in the city of Pittsburgh, was because of the result obtained with my mother. She had asthma for many years, fairly well-controlled by expert treatment.

However, in early summer, 1937, she started with uncontrollable attacks which lasted for over three months. Typically, she would go to bed around midnight, sleep for maybe one to one and a half hours and then awaken with severe wheezing and coughing, with great shortness of breath and no expectoration, and spent the rest of the time sitting up for a measure of relief. Mother had ultraviolet blood irradiation treatment in late September, 1937. That same evening she slept for five hours and, when she did awaken, she was able to expectorate bronchial exudate. She had three treatments, a week apart. Within about ten days the asthma was completely gone and has not returned.

Section 2

The Wide World
of
Photoluminescence

Chapter 7

The Effect of Hemo-Irradiation on Blood Poisoning

If we define septicemia [blood poisoning] as the presence of bacteria in the blood, the effects of photoluminescence on septicemia are key to innumerable treatment benefits. Since there are so many sources of this illness, hemo-irradiation will cure many health problems (technically, any viral infection will cause viremia – a form of blood poisoning, though it is seldom considered as such). From an infected scratch or tooth extraction to gas gangrene in a serious wound, photoluminescence will help clear the blood of its bacterial and viral contaminants. Thus, conditions as different from one another as the flu and shrapnel wounds respond well to photoluminescent treatment.

The improvement following hemo-irradiation is often quite dramatic in cases of blood poisoning. Subjectively, improvement in the mental state and clarity of thinking and speaking can take place within a few minutes of hemo-irradiation treatment. By the following day, reduction in toxemia is usually very noticeable, and is associated with improvement in the sense of overall well-being. Objectively speaking, the improvement is manifested through a fall in temperature, diminution of the sedimentation rate [the time it takes for red blood cells to settle in a test tube] and improvement in the general blood picture.

With a reduction in the infective process, there is a rapid fall in the white blood cell count back to normal. If the count has been low due to poor response to the infection, it frequently rises to an appropriate level for the severity of the infection. Then, as the process continues, the white count will return to the normal level again. Interestingly, photoluminescent treatment will frequently correct a low white count even when the condition is caused by certain drugs.[1]

Improvement in Peripheral Circulation in Treatment of Toxemia

In cyanotic individuals (those who are bluish in color from a lack of oxygen), a return to normal color often occurs even before the completion of the hemo-irradiation treatment.

Barrett reported the case of a young woman who was being treated for septic toxemia and had been under an oxygen tent for a number of days, with no apparent effect. Hemo-irradiation immediately returned her color to a healthy pink.

Patients usually report an increase in warmth of their extremities following hemo-irradiation. This improvement will often last for many months.

Relief of Pain

Patients with septic toxemia [toxic symptoms due to a bacterial blood stream infection] often complain of headache, muscular ache and other types of pain. However, when these patients are hemo-irradiated, their pain is typically relieved in a very dramatic manner. While the treatment takes only ten to fifteen minutes, the patient is often completely pain-free even before this short time lapses. Barrett commented that he thought this was a psychological effect, but it occurred so many times and with such almost immediate relief, that he could no longer explain it on that basis.

Toxemia due to an abscessed tooth will respond almost immediately to hemo-irradiation. Because the infection is walled off, however, this improvement is only temporary and extraction or root canal on the tooth is usually necessary.

Septic Arthritis with Night Sweats

Dr. Hunter, of Flushing, New York, reported the following case in 1938:

---

1 Aitken, *Ultra Violet Radiations and Their Uses*, Edinburgh, Scotland: Oliver and Boyd, 1930, p. 208.

This patient had rheumatic fever, which left her joints so painful she could scarcely walk. She seemed to recover completely, including her rheumatic heart condition. However, during the winter of 1937 and '38, she developed a mild arthritis and endocarditis. She had very severe night sweats and her weight fell off to 70 pounds. A month before coming to see Dr. Hunter for an irradiation, she had had a tooth extracted. A blood culture revealed the presence of streptococcus hemolyticus bacteria, undoubtedly secondary to the tooth extraction. This bloodstream infection had caused a recrudescence of her rheumatic fever.

At the time of hemo-irradiation the patient looked weak and ill. She had been sent to me to see if irradiations would have any effect on the uncontrolled night sweats. She had been under the care of several physicians for this, but no definite cause was found and all medications failed. The patient stated that she had to remove or change the bed linen three and four times nightly, summer and winter. She had, in addition to this, polyarthropathy (multiple joint pain and swelling), which was moderately severe, but it was because of the sweating that she had come to the author of this report.

The patient was given five irradiations in all. The first treatment was given on Saturday, June 24, 1938. On Sunday morning the patient phoned to say that she had perspired only about half the usual amount and that she felt better. The perspiring diminished more the following night and became progressively less. When she came for her second irradiation on the 15th of July, she was generally much improved, although the sweating was increasing somewhat. The second and subsequent irradiations were very effective and, following the third, she was completely over her sweating and it never returned.

Toxemia of Pregnancy

Toxemia is occasionally associated with pregnancy, though such occurrences are now rare. The cause is unknown, but the symptoms include high blood pressure and, occasionally, convulsions. Dr. Hunter reported on such a case:

A white female, age 23, became pregnant about November 15, 1937. Shortly after this she developed nausea and vomiting, and for about three weeks she could barely keep anything but water and fruit juices on her stomach. She lost about 24 pounds in weight in a period of about six weeks. She was very weak and her skin was covered with a great number of small pimples. She then started to improve, but at no time did she feel like herself. At six and one-half months she had a sudden eclamptic seizure after complaining of a headache the morning of the attack. At the hospital she was placed on sedation to which she responded fairly well. However, in a week's time the fetal heart sounds ceased and the fetus was removed. Following her return home, her weight was only 106 pounds (normal 128 pounds). It was decided to give her a hemo-irradiation and this was followed by a remarkable improvement in her health. She gained 36 pounds in seven weeks.

After an interval of two years she became pregnant a second time, but again she became very toxic and started to lose weight quickly. An irradiation was given with a very prompt improvement in her health and strength. Another irradiation was given in ten days followed by a third in ten days. She was given a prophylactic hemo-irradiation seven days, and another seven hours, before being delivered of a healthy child. A few weeks ago she became pregnant for a third time and, soon after, started vomiting and appeared to be about to have a repetition of previous toxemias. This time hemo-irradiation was given at once followed by a second in two weeks, and at the present time she enjoys much better health than in the earlier weeks of her previous pregnancies. She is maintaining her weight and has no signs of nausea or evidence of toxemia.

It would appear from these repeated bouts of toxemia that this is a superb prophylaxis for women suffering from this complication of pregnancy. In both episodes, treatment led to a healthy mother and a healthy child. Hochenbichler treated 100 patients with toxemia of pregnancy, even after the onset of convulsions, and none developed serious complications. All cleared completely with hemo-irradiation.

Septic Abortion

Another case of septicemia which "survived antibiotic therapy" was reported from the City Hospital in Welfare Island, New York:

A nineteen-year-old white female was admitted for vaginal bleeding on November 21, 1939. She stated that in February, 1939, she was delivered of a seven-month living premature infant by forceps.[2]

---

2 It is hard to imagine why forceps had to be used on a seven-month premature baby. This may have contributed to her subsequent serious complications.

Her first menstrual period returned on October 12, eight months after her delivery. Following that menses, she thought she was pregnant again. She pushed her finger into her uterus on November 12 in an attempt to abort the fetus and soon began bleeding with passage of clots. She continued bleeding until admission and, on the morning of admission, had a severe chill. She appeared acutely ill and her temperature was 100.8°F., pulse of 104, respiration 24....

The temperature ranged from 99.0° to 102.6°F. for the first two days. Cervical and urethral smears were positive for gonococcus [gonorrhea] and the patient was placed on sulfanilamide [sulfa] on November 25. Her temperature remained at 99°F. for three days but on November 28, after she had received 180 grams of sulfanilamide, she became cyanotic and had a chill with a temperature of 103.6°F. The drug was discontinued. The patient then began running a septic temperature up to 105.6°F. Seventy grams of sulfanilamide were given on December 21 and then again discontinued.

On December 3 the patient was given a 500 ml. blood transfusion. Her temperature on December 4 in the afternoon was 106.4°F. following a severe chill. All blood cultures were negative.

At this time, because the antibiotic therapy was obviously not working, and, in fact, making her sicker, hemo-irradiation was ordered and 130 cc of the patient's blood was irradiated and then reinjected. Her temperature rapidly fell to normal and remained so. She rapidly improved and was discharged in a few days completely without symptoms.

Appendicial Abscess

The following case was reported from New York City. As will be apparent, the patient may very well have died without the benefit of ultraviolet irradiation therapy.

E. M. was admitted to the hospital on September 19, 1941, because of pain on defecation, vaginal bleeding, and a tender mass in her abdomen. She had been operated on three years previously for a ruptured appendix and had been in bad health ever since.

She was taken to the operating room and found to have pus pockets in both of her Fallopian tubes with chronic peritonitis.

Fourteen days following the surgery, she developed symptoms again with pain in the lower abdomen and she started running a high temperature. A pelvic abscess was diagnosed and she was started on sulfa drug. This was discontinued the next day because she became cyanotic, presumably from the sulfa. Ten days later, a large abscess cavity on the right side was evacuated and 300 cc of pus was drained from the appendix stump.

She was again started on sulfa drugs and this time tolerated it well. But on September 20 her temperature went to 105°, so the sulfa was again discontinued. Because of the desperateness of the situation, hemo-irradiation was started, and, in 28 hours, the patient was completely well; her temperature was normal and her appetite was excellent. The draining sinuses over her appendix healed and closed completely. The patient gained 15 pounds in 15 days and was discharged perfectly well. She remained in excellent health.

Dr. Barrett reported an interesting "hopeless case" of bronchopneumonia in which antibiotics had completely failed to help the patient:

> I was called to treat a 25-year-old female, a relative of a physician, suffering from bronchial pneumonia. She had been taken ill 12 days previously and when seen by me she was in an oxygen tent. X-rays revealed characteristic bronchopneumonia in both lungs. The patient was moderately cyanotic. She was not responding to the sulfa drug therapy. Her physician and three consultants regarded the case as hopeless, and hemo-irradiation was requested as a last resort.
>
> The usual dosage of hemo-irradiation was employed in this case, and I requested that the oxygen be cut down considerably following the irradiation, for we have found that even before an irradiation is completed, the color usually improves due to greater oxygen absorption. It is not an uncommon experience to see the patient with cyanosis become pink even before the irradiation is completed. The following day this patient was receiving no oxygen. She was much stronger, and was sitting up. She convalesced satisfactorily having received only one hemo-irradiation.

In the preceding case, deemed to be hopeless by four different physicians, this response to only one treatment with ultraviolet light can only be described as miraculous.

Infectious Arthritis

Among his other important cases, Dr. Barrett described an instance of infectious arthritis with severe toxemia which was completely resistant to antibiotic therapy:

A male tool-and-die maker became ill on November 30, 1941 with severe pains and swelling in his left wrist. He was admitted to the hospital and 18 teeth were extracted in an attempt to solve his problem. His wrist became more swollen and painful and the effect of the extraction of his teeth on his infection was nil.

He remained in the hospital for 14 days during which time he lost 16 pounds. His physician requested hemo-irradiation as he was gradually getting worse, and when treated by Dr. Barrett on December 17, he was very toxic and his hand was greatly swollen and tender, so much so that he could not bear to have anyone touch it.

Hemo-irradiation was started and within one hour after the irradiation he could move his fingers and permit examination of the hand and wrist without discomfort. He had a second irradiation the following day and was discharged and returned to work and has been well since.

Septicemia with Cerebellar Thrombosis

If there was ever a hopeless case, it was one reported by Dr. Barrett of the brother-in-law of a New York physician. While vacationing in Miami Beach, the gentleman fell seriously ill. Dr. Barrett was called to Miami from New York to treat him because the situation was considered hopeless. Since nothing short of a miracle would help him, the attending physicians decided to try hemo-irradiation.

The patient had a thrombosis of the cerebellar artery (blood clot in the brain), pneumonia, a bacterial infection of the blood, emboli (clots) of the lungs, a blood clot in the major vein of his left leg, a paralysis of the left side of his body and a paralysis of the left vocal cords – a hopeless and certainly terminal case.

When seen by Dr. Barrett, the patient was delirious and irrational. He had eaten nothing except Coca Cola for 11 days and had lost 45 pounds.

He was immediately treated with hemo-irradiation and had an almost instant response. After a second treatment in three days, there was further dramatic improvement. Although it took him several months, he recovered completely, gaining back his lost 45 pounds and adding on another ten. Dr. Barrett reported:

> This case is reported to draw attention to the practically hopeless condition of a patient suffering from what was

discovered in New York to be a thrombosis of the cerebellar artery with many other complications. The hemo-irradiation was given in the hope of controlling the marked toxemia (he had, on a number of occasions, a sed rate of 120) and also to build up his resistance. During his hospitalization in New York, he developed a thrombosis of the femoral vein on the left side (in the leg) accompanied by a rise in temperature and very severe pain. Various measures were used without avail, but after one irradiation there was a prompt response and his complication disappeared entirely in three days. In all, this patient was given nine hemo-irradiations, three in Miami, three at the Manhattan Eye, Ear and Throat Hospital, and three after he returned home. During his convalescence at home, it was discovered that he had a toxic goiter (thyroid) which accounted for the continued elevation of his pulse rate.

This patient would almost certainly have died, probably within a few days, without the hemo-irradiation therapy of Dr. Barrett. In a five-year period from 1938 to 1943, Barrett treated over 400 patients suffering from a variety of conditions, infections of various kinds, systemic and regional, including infections of the eye. He used the ultraviolet method in about 30 cases of asthma and in about 60 cases of arthritis. In all, he treated about 35 different diseases.

Although external ultraviolet spectral energy had long since proved to be a therapeutic agent in a variety of conditions, such as rickets, infantile tetany, extra-pulmonary tuberculosis, lupus vulgaris and various other skin conditions, the introduction of ultraviolet irradiation of the blood (rather than the skin) dramatically increased the scope of therapeutic usefulness of ultraviolet. Barrett and others have proved it extremely useful in bacterial infections, toxemias, viral infections, and other serious infective conditions in which antibiotics often proved to be ineffective.

As Barrett put it:

> It is a method which raises the general resistance of the individual, diminishes toxemia rapidly and stimulates the healing forces of the body. It is a safe method with no untoward reactions. It can be used to supplement other recognized measures, operative and medical, but in its own right it is a measure which will often affect a cure when all other therapies have failed.

Some Typical Cases of Septicemia *

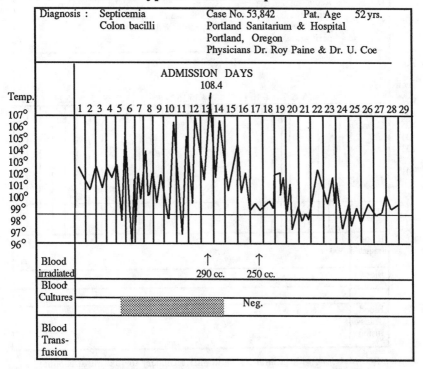

Diagnosis :	Septicemia	Case No. 53,842	Pat. Age	52 yrs.
	Colon bacilli	Portland Sanitarium & Hospital		
		Portland, Oregon		
		Physicians Dr. Roy Paine & Dr. U. Coe		

Fig.53842. Portland Sanitarium and Hospital. In this case of colon-bacillus septicemia secondary to acute pyelitis, blood-irradiation therapy was instituted when the patient's blood cultures had been positive for thirteen days; she had been in a coma for 40 hours, and her temperature had reached 108.4. Forty-eight hours after the initial blood irradiation the cultures became negative and the toxic symptoms began to subside; 96 hours later the patient's condition was greatly improved, the temperature falling to normal. A second blood irradiation was given as a precautionary measure. She convalesced quite well, leaving the hospital in apparently excellent condition fifteen days after the initial irradiation.

---

* Rebbeck and Miley, *Review of Gastroenterology*, Jan.-Feb., 1943.

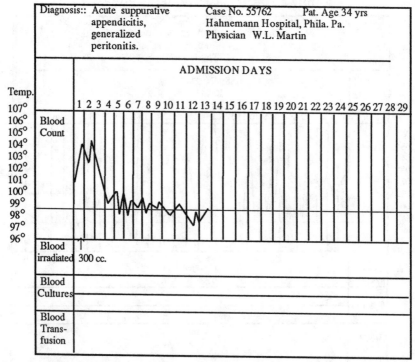

Diagnosis:: Acute suppurative appendicitis, generalized peritonitis.	Case No. 55762 Pat. Age 34 yrs Hahnemann Hospital, Phila. Pa. Physician W.L. Martin

Fig. - S83105. Male, 34, with generalized peritonitis secondary to rupture of suppurating appendix. Ultraviolet blood-irradiation therapy was given on the first postoperative day. Within 48 hours a subsidence of the classic symptoms of toxemia due to generalized peritonitis — namely, rapid, bounding pulse, rapid respirations, abdominal pain, distention and rigidity, Hippocratic facies, and irritability and apprehension— was clearly evident. The patient convalesced uneventfully thereafter, leaving the hospital in apparently excellent condition ten days after this single blood irradiation.

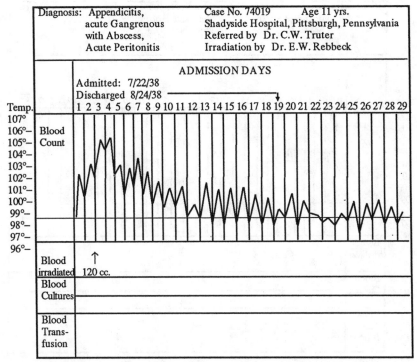

| Diagnosis: Appendicitis, acute Gangrenous with Abscess, Acute Peritonitis | Case No. 74019   Age 11 yrs. Shadyside Hospital, Pittsburgh, Pennsylvania Referred by Dr. C.W. Truter Irradiation by Dr. E.W. Rebbeck |

Fig. - H55762. Female, 11, with generalized peritonitis and appendiceal abcess secondary to perforation of a gangrenous appendix. Ultraviolet blood-irradiation therapy was given the first postoperative day. Her advanced toxic symptoms began to subside in 48 hours and detoxification was marked at the end of 72 hours; her temperature remained elevated, but continued to fall by lysis. She was discharged on the eighteenth postoperative day. This record shows daily peak temperatures within normal limits for an additional 10 days.

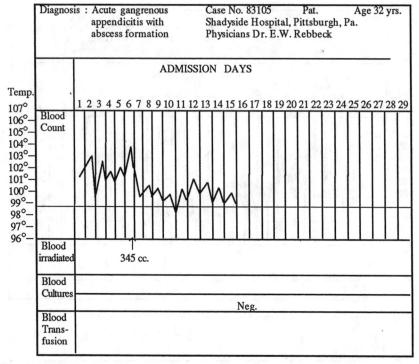

Fig. - S74019. Male, 32, with acute gangrenous appendicitis with abcess formation. After appendectomy the patient's condition was good for two days, but on the third postoperative day after complaining of increasingly severe headache, he suddenly and unexpectedly fell into a stupor. The next day his temperature rose to 104 degrees, despite a soft abdomen. He became unexplainably comatose and his condition seemed most grave. Ultraviolet blood-irradiation therapy was given; 24 hours later the stupor and headache disappeared and the temperature fell to 99.6 degrees. From this point on his convalescence was uneventful, and hospital discharge occurred nine days later.

Chapter 8

A Bacterial Mystery Finally Solved

Many of the original investigators of photoluminescent therapy have found that the treatment for bacterial infections was consistently excellent except in the case of Staphylococcus aureus infection. In 1943, Dr. Henry A. Barrett made the observation that hemo-irradiation of the blood, as it was called in those days, was being "largely eclipsed by the introduction of sulfa drugs...."

No one at the time realized that there would be any relation between the use of sulfa drugs and the inability of photoluminescence to affect Staphylococcus infections. However, Dr. George Miley did note in 1942 that the sulfa drugs could not be given safely within the first five days following ultraviolet blood irradiation therapy without risking a severe toxemia, pulmonary edema (excessive serous fluid in the lungs) and renal (kidney) shutdown. He also noted that those patients who have photo-oxidation therapy have a much shorter convalescent period than those who receive *both* sulfa drugs *and* photo-oxidation therapy.

The situation was quite confusing because some patients with Staphylococcus septicemia survived and had no more stormy course than those with other infections, but others promptly died. Virgil K. Hancock reported on two cases of Staphylococcus septicemia, which is generally fatal, who recovered promptly with hemo-irradiation.[1] A study of Hancock's patients revealed that neither of them received

---

1 Rebbeck, *American Journal of Surgery*, Vol. 63, No. 3, December 1942.

any sulfa drug before, during, or after his treatment with photo-irradiation.

Building on the work of these pioneers, Dr. George Miley finally solved the puzzle as to why Staphylococcus septicemia did not respond to photo-illumination.

In his landmark article in the *American Journal of Surgery*, June, 1944, Miley stated:

> Our findings have been in close agreement with the other workers, and have been reported as such with one exception, mainly, the value of ultraviolet blood irradiation therapy in the treatment of Staphylococcus septicemia, in which we originally reported the failure of ultraviolet blood irradiation therapy to control the progress of Staphylococcus septicemia in seven consecutive cases. Recently, we have observed complete recovery from Staphylococcus septicemia in nine consecutive cases, *and therefore wish to retract our original statement that we believe Staphylococcus septicemia did not respond well to ultraviolet blood irradiation therapy.*

Miley noted that Dr. Hancock had reported eight patients suffering from Staphylococcus septicemia who received no sulfa drug therapy and recovered with ultraviolet blood irradiation therapy alone.[2]

Miley reported:

> As the result of our original experience of seven consecutive failures, followed by nine consecutive recoveries of Staphylococcus septicemia, following ultraviolet blood irradiation, we naturally wished to find some reason for the marked difference in results and, in carefully going over the records of all sixteen individuals, *we found that in six out of seven cases of the original failure group intensive sulfa drug therapy had been given before and / or after ultraviolet blood irradiation therapy was administered, whereas eight of the nine individuals in the recovery group had had no sulfa drugs whatsoever....*

The following table reveals the dramatic difference between those treated with sulfa drugs and those untreated:[3]

---

2 Hancock and Knott, *Northwest Medical Journal*, June 1934.
3 Miley, *The American Journal of Surgery*, June 1944.

Effect of Sulfa Therapy on Blood Irradiation

No.	Hospital No.	Type of Staphy-lococcemia	Primary Infection	Sulfa Rx	No. of Blood Irradiations	Result
1	52117	Aureus	Staph, Pneumonia, lung abscess	Yes	2	Died
2	n/a	Aureus	Unknown	Yes	2	Died
3	65236	Aureus/Albus	Puncture wound of eye	Yes	4	Died
4	64501	Aureus	Prostatic resection area infection	Yes	1	Died
5	48830	Aureus	Wound infection, sinus thrombosis following frontal sinusectomy	Yes	1	Died
6	60720	Aureus	Wound infection following operation for ingrown toenail	Yes	2	Died
7	38082	Aureus	Bladder carcinoma, bilateral pyonephrosis, emphysema, atelectasis, bronchial pneumonia	None	4	Died
8	81994	Aureus	Marked erysipeloid inflammatory process of right ear	None	1	Recovered
9	84630	Aureus	Incomplete septic abortion	None	1	Recovered
10	88168	Aureus	Incomplete septic abortion	None	2	Recovered
11	88167	Aureus	Incomplete septic abortion	None	1	Recovered
12	82484	Aureus	Incomplete septic abortion	None	1	Recovered
13	83141	Albus	Acute Ulcerative rhinitis, acute suppurative otitis media, acute mastoiditis, incomplete septic abortion	None	2	Recovered
14	82702	Albus	Incomplete septic abortion, putric endometritis, parametritis, pelvic peritonitis	None	2	Recovered
15	86768	Aureus	Post-measles upper respiratory infection	None	2	Recovered
16	50698	Albus	Post-caesarean pelvic thrombophlebitis	None	1	Recovered

Miley made the following clinical observations:

> The recovery of nine individuals with Staphylococcus septicemia receiving ultraviolet blood irradiation alone were certainly opposite from the results observed in the seven individuals who also were suffering from Staphylococcus septicemia but who died in spite of ultraviolet blood irradiation therapy given after a long and intensive sulfa drug therapy. The chief difference between the two groups receiving ultraviolet blood irradiation therapy is obviously the use of the sulfa drugs in the group that died.

In the recovery group, there occurred the same sequence of events already reported to occur in other acute pyogenic infections. These were:

1. A marked detoxification effect.
2. A complete disappearance of the invading bacteria organism.
3. A grossly discernible peripheral vasodilatation.
4. A complete absence of deleterious effects.

> *Apparently the use of sulfa drugs so greatly lowered the resistance of the individuals ... that ultraviolet blood irradiation therapy was no longer able to be of any benefit to these patients....* Once sulfa drugs fail to control an infection one encounters all the ill effects of a toxic drug, producing generalized tissue anoxia, superimposed upon the toxic products of bacterial growth and decomposition.... It is our opinion that the use of the sulfa drugs in the cases of Staphylococcus septicemia seen by us has had no beneficial effect whatsoever, but on the contrary, has so seriously lowered the individual patient's resistance that recovery was made impossible.

Miley's final conclusion:

> We must retract our original statement that ultraviolet blood irradiation therapy has little or no affect on Staphylococcus septicemia, and admit that, when used alone, this procedure has been successful in controlling Staphylococcus septicemia, at least in nine consecutive cases.

Chapter 9

Thrombophlebitis

The sudden appearance of acute thrombophlebitis[1] is one of the most unpredictable, dangerous, and undesirable complications encountered in medicine. It is also a complication of surgery that can appear apparently without cause. In many ways, it remains "one of those mysteries of medicine." Various methods come in and out of fashion trying to control this condition satisfactorily, but their failure to do so is well known. Now, however, we have a treatment in photoluminescence that targets this severe disease perfectly.

Miaui reported five cases of acute thrombophlebitis with a *recovery rate of 100 percent*. He noted a rapid disappearance of pain and tenderness, usually within 24 to 48 hours. All five of his cases were antibiotic failures.

In 1943, Dr. George Miley reported on 13 consecutive cases of thrombophlebitis. He found that photoluminescence controlled the classic symptoms of acute thrombophlebitis rapidly and efficiently in every case treated. This included thrombophlebitis with or without a fever which followed normal natal delivery, operations, or acute pyogenic infections.

A study of Miley's table of cases shows a definite uniformity of response to phototherapy. Pain and tenderness are almost always the first symptoms to disappear, followed by fever, if present. Edema (fluid in the tissues) is the last of the signs to subside, often taking a number of weeks.

Five of the cases he listed entailed the failure of chemotherapy, bed rest, heat, and elevation prior to photoluminescent therapy. All these cases responded almost immediately to phototherapy,

---

1 Inflammation and infection of the veins, usually in the lower extremities.

usually in just one treatment. In every case, without exception, pain, tenderness and fever disappeared within 24 to 48 hours.

Even Dr. Miley's worst case presents a clear illustration of the remarkable therapeutic effects of extracorporeal ultraviolet irradiation on this formerly difficult-to-treat disease.

A 48-year-old female was admitted to the hospital with a five-day history of nausea, vomiting, and a spreading infection of the right groin. The infection apparently originated from an infected varicose ulcer located along the course of the internal saphenous vein (inside the thigh). Physical examination revealed thrombophlebitis of the femoral vein with marked hardening of the lymph glands in the groin, and a spreading erysipeloid[2] lesion extending from the upper groin region down the inner surface of the thigh, halfway to the knee. The lesion was discharging an extremely foul-smelling fluid from several areas, and much of the adjacent skin area was gangrenous.

Her temperature on admission was 100.8°F.; her pulse was 120. She had a white blood cell count which indicated bacterial infection and she was extremely toxic and debilitated.

At the end of 48 hours, she was worse, with a temperature of 103.4°F., at which time ultraviolet blood irradiation was instituted. Within the first 24 hours after treatment, the patient's toxic symptoms began to subside, the groin lesion broke down in several places, and the foul discharge became more profuse. On the third post-irradiation day, the temperature began to drop further and the drainage was markedly less. The patient's general condition had obviously improved markedly. She continued to improve, with the inflammation of the femoral vein subsiding and a complete disappearance of pain, tenderness and swelling within five days. After a total of eleven days, and a single blood irradiation, she was in excellent general health, with complete healing of the skin, and was discharged to her home.

The patient could clearly have died without use of photoluminescent therapy. In spite of the startling results reported by Miley and others, this treatment of thrombophlebitis is not available in a single hospital in the United States.

---

2 Intense, local redness of the skin with swelling.

As an historical aside, "therapeutic nerve blocks" were in vogue during the '40s for the treatment of thrombophlebitis but were, in most cases, entirely ineffective.

Miley made the interesting observation that photoluminescence "probably does not exert any prophylactic protective effect." In one of his cases, there was a new infection two weeks after a previous one had subsided. He said, "The fact that the newly-occurring thrombophlebitis in turn subsided rapidly after ultraviolet irradiation therapy, and did not recur, speaks for itself." This brings up the question of whether photoluminescence is prophylactic for *any* disease process, but the answer is not known.

There are four basic biochemical and physiological effects of ultraviolet light therapy: bactericidal activity, detoxification, vasodilatation and increased blood oxygen values. These effects are well-known and have been adequately reported. It seems unlikely, Miley remarked, that any of these four effects of ultraviolet could alone or combined completely account for the remarkable results seen in the treatment of thrombophlebitis. It seems that one or more physiological, neurological, or biochemical effects that are not understood must be taking place to account for the profound and rapid change seen in the treatment of acute thrombophlebitis with ultraviolet irradiation.

It is interesting to note that ultraviolet blood irradiation does not cause an increase in the tone of intestinal muscle in a normal person; there are no abdominal cramps or other indications of a change in the physiological status of normal intestinal musculature. Yet, when photoluminescence is given to a patient with an obviously disturbed autonomic nervous system involving smooth muscle tone of the gastrointestinal tract, this part of the autonomic nervous system, hopelessly out of balance, comes very quickly back to normal. This return to normal muscular tone obviously occurs in the treatment of thrombophlebitis just as it does in a gastrointestinal smooth muscle disease. Somehow, the smooth muscle elements of that portion of the peripheral vascular system that are diseased are directly affected by the energized light; the rest of the system is not. In view of the fact that the photon energy attacks only the localized peripheral autonomic vascular imbalance, it would appear the body sets up some sort of temporary omni range, that brings the necessary photonic energy to bear on the specific area where it is needed. There is

Results of Ultraviolet Blood Irradiation Therapy in Thrombophlebitis*

No.	Age and Sex	Veins In- volved	Etiological	Symptoms Present at Time of Initial Blood Irradiation				Results	No. of **UVBIT	No. of Days in Hospital after initial **UVBIT
				Edema	Pain	Ten- der- ness	Temp. F.			
1	40 F	LF	Postoperative (nephropexy)	+	+	+	99-104.4	Disappearance of pain, tenderness, and fever in 24 hrs., of swelling in 7 days.	1	11
2	30 F	LF	Postoperative (appendiceal abscess)	+	+	+	98-100.8	Disappearance of pain, tenderness, fever and swelling in 48 hrs.	2	11
3	33 F	LF	Postoperative	+	+	+	None	Disappearance of swell- ing, pain, and tenderness in 48-72 hours.	1	14
4	42 F	LF	Postoperative (hysterectomy)	+	+	+	99-101.6	Definite reduction of pain, tenderness, and swelling in 48 hrs., temperature falling to normal. All disappeared in 7 days.	1	7
5	32 F	LF	Postpartum	+	+	+	103.8	All symptoms began to subside in 48 hours.	1	22
6	18 F	LF	Post-caesarean Staph albus septicemia	None	+	None	102.0	Leg pain disappeared in 4 hrs., temperature nor- mal after 2nd irradiation.	2	11
7	48 F	RF RS	Postinfectious (infected varicose ulcer, erysipeloid skin lesion, lymphadenitis)	+	+	+	103.4	Pain, tenderness, edema disappeared in 5 days; skin in- fection also subsided in 5 days.	1	11

** Ultraviolet Blood Irradiation Therapy

No.	Age/Sex	Vessel	Diagnosis			Temp	Course		Days
8	43 M	LF	Postinfectious, wound of foot	+ present 3 months, indurated	+ +	None	In 48 hrs. pain and tenderness subsided, but not edema; able to walk for first time, since onset of pathology.	1	8
9	24 F	LF RF	Postpartum	+	+ +	101-102	Pain, tenderness, swelling and fever (recurrence) disappeared in 12 hrs. Patient readmitted with feverless recurrence which subsided completely in 14 days; no further recurrence.	1	15 14
10	47 F	LF	Postoperative (hysterectomy)	+	+ +	102	Pain, tenderness, and swelling gone in 5 days; temperature fell to normal on 6th post-irradiation day, after removal of heat cradle.	1	18
11	65 M	LF	Postinfectious (severe dental caries)	+	+ +	None	All signs of phlebitis disappeared in 15 days; teeth extracted, renal colic present, in remaining 13 days.	1	28
12	18 F	RS	Postpartum	+	+ +	99.6	Pain, tenderness gone in 10 days; fever and swelling in 72 hours.	1	7
13	32 F	RF LF	Postpartum (also pulmonary embolism)	+	+ +	103.8	Leg and chest pain, tenderness and fever disappeared in 48 hours; swelling in 4 days.	1	17

Key LF—left femoral.   RS—right saphenous.
RF—right femoral.   LS—left saphenous.

*Miley, *American Journal of Surgery*, June, 1943.

a marked, generalized peripheral vasodilatation (a pinkening of the skin) with, as Miley says, "almost monotonous frequency" following a blood irradiation. However, in the diseased blood vessels and the musculature of the intestine, the effect is exactly opposite. The effect there is a *vasoconstriction* – a constricting of the veins.

Miley summarized this mysterious action as follows:

> It is interesting to observe that the grossly discernable vasodilatation effects noticed following application of ultraviolet blood irradiation therapy occur in apparently normal humans, as well as in those who are suffering from various disease processes, whereas an increase in tone of normal gastrointestinal smooth muscle is not apparent, at least not to the extent that it produces colicky or cramp-like contractions. Likewise, as just stated, no evidence of an increase in smooth muscle tone, as in acute thrombophlebitis with edema, appears in normal individuals following blood irradiation but, on the contrary, only vasodilatation occurs. Therefore, it must be assumed that there exists a profound and fundamental regulating effect of the radiated blood upon the autonomic nervous system; this effect could be described as a 'normalizing effect,' and it becomes most apparent in definite imbalances of the autonomic nervous system.

It is as though the photon-activated blood has a life of its own and knows where to go – homing in on diseased veins for instance; and what to do – increasing the tone of the diseased vessels; whereas there is a generalized vasodilatation in the rest of the body.

A less sophisticated theoretical explanation, Miley points out, would be that individuals who respond to ultraviolet blood irradiation therapy so dramatically may be suffering from ultraviolet deprivation brought about by (1) a breakdown of the body's mechanism for absorbing ultraviolet, (2) inadequate supply of ultraviolet, or (3) a combination of both.

Chapter 10

Bronchial Asthma

In reading through this book and noting the many diseases and conditions photoluminescence treats effectively, it's easy to see that ultraviolet treatment is death on viral contaminants in the body. However, you wouldn't expect it to be effective in something like bronchial asthma, since asthma is thought to be primarily an allergic disease;[1] any infection involved is generally considered secondary. Nevertheless, there is a group of asthmatics whose condition derives from an allergic response to their own infection. These cases would be expected to improve with ultraviolet therapy, and indeed they do.

Dr. George P. Miley, et al, of Philadelphia, Pennsylvania, studied 80 persons suffering from severe intractable bronchial asthma. It is important to note that these patients were *intractable,* meaning no therapy was helping to any material extent. This study was conducted from November 1, 1938 to October 1, 1942. "Intractable," in these cases was defined as:

1. Failure to respond to removal of known aggravating extrinsic factors such as dust, pollens, foods, etc.
2. Unresponsiveness to all desensitization efforts.
3. Unresponsiveness to bronchoscopy and bronchoscopic drainage.
4. Failure to respond to sinus infection therapy.
5. Temporary or no relief following inhalation or injection of epinephrine (an adrenal hormone) or epinephrine-like preparations.

---

1 At the end of this chapter we will give you some updated information on how light therapy does actually work directly on the allergic process, not just through clearing an infectious process.

The ultraviolet blood irradiation treatments were administered every four to six weeks until the patients' symptoms had improved. After a few months, the frequency was reduced to a treatment every eight to ten weeks.

Injections of epinephrine and other medications were withheld as much as possible. In most cases, they were gradually and then completely eliminated. The overall results were described by the authors as "extremely gratifying." The following case histories are illustrative:

* * * * *

Mrs. A. S., a 46-year-old female, was admitted to Hahnemann Hospital with intractable bronchial asthma and chronic sinusitis, which she had had for the previous 26 years. Her asthma had been unrelieved by de-sensitization, bronchoscopy, vaccine therapy, various intra-nasal operations, cholecystectomy (gall bladder removal), and injections and inhalation of epinephrine and epinephrine-like products.

Initial physical examination revealed a severely debilitated patient with marked generalized cyanosis (blueness of the skin due to poor oxygen absorption), and extreme weakness. She was, according to the examining physician, "suffering a severe psychoneurosis secondary to her ill health."

Ultraviolet blood irradiation therapy was administered the same day of admission. This was the only treatment she received during her admission. The rest of her therapy consisted of the traditional medications for severe asthma. She was discharged from the hospital in four days.

The patient was next seen a month later when she reported that she had suffered several asthmatic attacks since the first blood irradiation, but stated they were not so severe or so frequent as previously. Blood irradiation therapy was repeated at this time.

She was next seen three months later, at which time she reported her asthmatic attacks were even less severe than in the first few months following initial blood irradiation, and she had gained seven pounds. She was next seen three months later when she again reported the attacks were steadily becoming less frequent and less severe.

Five months later, she was given another blood irradiation therapy and the doctors reported: "Her condition was obviously greatly improved, her attacks were much reduced in number and severity and it was noted that cyanosis had practically disappeared."

The therapy was repeated about every two months for the following six months and, when last seen on April 27, 1942, the doctors reported: "When last seen the patient was having only one to three almost negligible attacks per month. Her weight had risen from 102 to 122 pounds and her general condition permitted her to do all the work for a household of six as contrasted with her complete invalidism when she was first seen."

* * * * *

Mrs. M. M., age 56, was first seen on November 22, 1938. She gave a history of 17 years of constant, severe asthmatic attacks with insomnia the previous six years. The best types of treatment had been employed, notably desensitization to house dust, the only known predisposing etiologic factor, use of autogenous vaccine made from cultures of bronchial bacteria, a bronchoscopy, sinus extirpation [opening of the sinus with drainage], inhalation and injection of epinephrine and ephedrine preparations and inhalation of the smoke of burning stramonium leaves.

Despite these and other well-known procedures, the patient's general condition slowly but steadily deteriorated so that when first seen she presented the characteristic picture of advanced, intrinsic, and intractable bronchial asthma. She had marked and generalized cyanosis, insomnia, a definite inability to tolerate ephedrine or epinephrine in any form, marked cachexia (weight loss), eight percent eosinophilia, X-ray evidence of emphysema and, as might be expected, a psychoneurosis bordering on perpetual hysteria.

Photoluminescent therapy was administered on the day of admission. That night the patient experienced an aggravation of her asthmatic symptoms roughly equivalent to her average severe attack. However, the following day she improved slightly; this improvement continued steadily and, four days after blood irradiation, she was obviously better. Her attacks continued, but were much milder. A second blood irradiation was given two weeks later with no aggravation of her symptoms; only one attack occurred the following week and she slept well every night. She continued to improve, with her cyanosis beginning to disappear after another week, and her general condition was greatly improved. Dr. Miley remarked, "She no longer smokes stramonium leaf cigarettes at night."

The patient was discharged after receiving a total of 15 luminescent therapies during the next nine months. She was able to support herself and her husband by selling lace novelties. Her only severe recurrence was on July 7, 1941, after a seven-month withdrawal of all treatment. This gradually disappeared after three applications of ultraviolet irradiation therapy. She seldom had any problems following this, but when she did have a slight occurrence, it was easily controlled with the repetition of the therapy every three or four months.

* * * * *

J. M., a man of 21, was a remarkable case in that he responded to only one photoluminescent treatment. He had a history of hay fever attacks followed by bronchial asthma for three years. He was having one or two moderate attacks per month when seen and, after one ultraviolet blood irradiation, he had no further symptoms. Since he was last seen, 24 months previously, he had experienced no symptoms of asthma whatsoever. He was also free from hay fever.

Analysis of the Miley Cases

A resumé of Miley's cases shows that of the 56 persons receiving therapy, 45 were definitely improved. Nine of these were greatly improved and maintained an improvement for more than a year; 20 were moderately improved and maintained improvement for over six months. The 11 unimproved patients failed to show any change after therapy and after observation over a period of six to ten months. However, it should be noted, that in the early days of phototherapy, there were no photosensitive compounds to enhance the regimen, and there was no use of bio-oxidation (intravenous hydrogen peroxide), which, in our experience, greatly enhances the treatment.

Twenty-four of the patients did not receive sufficient treatment to permit any critical observation.

The following table analyzes the results in various types of asthma. As can be seen from this breakdown, there is a possibility that *any form* of allergy may respond to the therapy. (Although, it should also be noted, a fair percentage of patients did *not* respond.) The next table is a composite of all the bronchial asthma cases and, as can be seen, about eighty percent had good to excellent results.

Clinical Observations on
Asthma Therapy

Having thus treated quite a few asthma patients, several observations can be made:

1. During the first night after the initial blood irradiation, a moderately severe asthmatic attack is likely to occur, especially in patients with severe asthma.
2. Marked relief in the frequency and severity of the dyspnea attacks is noted after the second or third blood irradiation, given at intervals of three to six weeks after the initial treatment.
3. During the first year, it is advisable to administer ultraviolet blood irradiation therapy eight to twelve times at three to six week intervals.
4. After the first year, the repetition of blood irradiation two to five times yearly will usually keep a severely affected patient free from asthmatic attacks.
5. A cyanosis of many years standing slowly disappears within a year of initiation of photoluminescence.
6. A marked and relatively permanent increase in the patient's general resistance occurs, so much so that many patients who have been under constant treatment will be enabled to lead normal, useful lives again with little or no discomfort.
7. A six-month withdrawal of ultraviolet blood irradiation therapy after one or two years will occasionally cause a moderately severe exacerbation of the asthma. After irradiation is repeated, the symptoms again subside, though in some cases more slowly than originally.
8. Use of ephedrine or epinephrine inhalants is required only rarely for most patients.
9. There is a complete absence of any toxic effects from the treatment in all patients.

As with all new and innovative therapies, the critics claimed the effect was probably psychosomatic. This claim is always put forth by those who are unwilling to accept dramatic changes in modes of treatment either because of ego or a vested interest in an old established therapy. The doctors made some very interesting and somewhat biting comments at the conclusion of their study which, in effect, put the "psychosomaticists" in their place:

> Since it is obviously impossible to eliminate the psychosomatic effect of a new therapeutic agent on an asthmatic patient who has run the gamut of pseudo-scientific therapeutic measures, we have not attempted the impossible; however, in order for a therapy to be judged as purely psychotherapy, it

must be proved to have no therapeutic effect on any disease. But to anyone who has seen apparently moribund patients, in the last stages of acute pyogenic infection, recover dramatically after ultraviolet blood irradiation therapy, a purely psycho-therapeutic theory seems pitifully inadequate and wholly untenable. We are sure that, if any of us had septicemia, we would prefer blood irradiation to the best psychosomatic treatment by the best psychiatrist.

Nevertheless, the psychic element cannot be ignored in the evolution of any new therapy used in a chronic, debilitating disease; therefore, we have tried to base our evaluation or criterion of improvement on the length of time the patient's asthmatic symptoms have been controlled. If a patient has improved greatly for two months and then returns to his original asthmatic status, he is unchanged except for a temporary psychosomatic effect, an effect which characteristically wears off, as a rule, within a few months at the latest. If a patient has had intractable bronchial asthma for 38 years, plus insomnia for six years despite the best of medical care, and has little or no asthma, plus freedom from insomnia, for almost four years from the time she first is given ultraviolet blood irradiation therapy, she obviously has had much more than psychotherapy. If the same patient has had, unknown to her, a withdrawal of the therapy by closing off the hemo-irradiator, cutting off the source of ultraviolet rays from the returning blood, and begins to show signs of severe asthma again, she is suffering from a true ultraviolet deprivation. If we next really give the patient ultraviolet blood irradiation therapy and her symptoms once more clear up, we feel that this is proof positive that such a patient – and we had such a patient – not only has been benefitted by the ultraviolet irradiation of blood, but has been suffering for years from severe ultraviolet deprivation, which apparently plays a much greater part in this disease, as well as others, than has been suspected.[2]

Fifty years have passed since Miley and his colleagues attacked the foundations of modern medicine with photo-therapy. There is now a scientific explanation for the use of photo-irradiation in allergy. All toxic substances in the body have a fluorescence,

---

2 Miley, et al, *Archives of Physical Therapy*, September 1943.

including chemicals (tobacco), sick cells (cancer), and allergens (weeds, strawberries, or whatever you are allergic to). This fluorescence is, in effect, a marker which light energy targets for destruction. So when light energy (photons) are introduced directly into the blood, as in photo-oxidation, the recovery is often dramatic.

Results in Bronchial Asthma—Miley, et al*

Summary of Cases**

Case No. and Patient's Name	Sex & Age, Years	Duration of Disease Previous to Blood Irradiation, Years	Condition at Time of First Irradiation	No. of Irradiations	Mos. Under Treatment & Observation	Date of First Irradiation	Condition on 10/1/42	Result
1 F.V.R.	M 55	31	15-20 severe attacks monthly	6	36	10/17/39	3-5 moderate attacks monthly	xx
2 G.U.	F 61	31	10-15 moderate attacks monthly	4	16	5/23/41	5-10 mild attacks monthly, general condition improved	x
3 D.E.S.	F 30	28	25-40 severe attacks monthly	4	41	5/1/39	1-2 mild attacks yearly	xxx
4 A.S.	F 46	26	Extreme emaciation and cyanosis, 40-50 severe attacks monthly	9	24	5/17/40	1-3 mild attacks monthly, no cyanosis, 20 lb. increase in weight	xxx
5 B.C.	F 24	24	30-50 attacks monthly since infancy, marked cyanosis	4	15	7/11/41	Free of attacks 11 months, for first time in her life, aggravation last 6 weeks	xx
6 G.C.	M 49	19	30-50 attacks monthly	5	23	11/12/40	1 mild attack monthly	xxx
7 M.M.	F 56	17	Severe asthma, insomnia for past 6 years, cyanosis	17	46	11/22/35	1-2 mild attacks monthly, no cyanosis, one seasonal aggravation	xxx

8 W.J.	M 50	15	15-20 severe attacks monthly associated *with pansinusitis, cyanosis*	8	3/18/40	5-15 attacks monthly, cyanosis	0
9 F.D.	F 30	15	20-30 attacks monthly	1	9/6/40	Unimproved	
10 F.J.	M 30	15	1 severe attack lasting 3 to 7 days, monthly	2	5/17/40	Excellent, no monthly attacks	xxx
11 J.D.	F 14	11 1/2	1 severe and 3-4 mild attacks monthly	6	11/15/41	No attacks as yet	xx
12 E.L.	F 54	11	30-50 severe attacks monthly	2	6/29/39	Not seen since	N.F.
13 R.M.	F 27	10	In oxygen tent owing to constant severe attacks (status asthmaticus)	2	9/9/40	Discharged improved, attacks easily controlled by spary (vaponephrin) aponephr inhalant up to time of discharge from hospital	N.F.X.
14 V.L.	F 42	8	Severe attacks every 8-9 hours	4	11/18/41	5-7 mild attacks monthly	xx
15 A.S.	M 34	8	30-50 attacks monthly with constant insomnia	3	9/13/41	Unchanged	0
16 C.D.	F 53	6	Two severe attacks in 3 months	3	10/20/41	No attacks	xx

*Archives of Physical Therapy, September, 1943.
**See last page of table for key

Continued from page 125

Case No. and Patient's Name	Sex & Age, Years	Duration of Disease Previous to Blood Irradiation, Years	Condition at Time of First Irradiation	No. of Irradiations	Mos. Under Treatment & Observation	Date of First Irradiation	Condition on 10/1/42	Result
17 C.S.	F 22	6	10-20 attacks monthly	14	13	8/30/41	No attacks in 11 months, seasonal aggravation in last 2 months	xx
18 A.G.	M 37	4	Constant wheezing, daily mild attacks	5	13	11/13/40	Unchanged	N.F.
19 A.M.	F 41	6	Status asthmaticus	1	less than one	9/12/41	....	N.F.
20 J.E.	F 43	4	Severe attacks during spring and fall despite desensitization	2	25	9/14/40	Spring and fall attacks disappeared	xxx
21 J.M.	M 21	3	1-2 mild attacks monthly	1	25	9/11/40	Complete absence of attacks since first irradiation	xxx
22 R.B.	M 14	2 1/2	Severe monthly attacks at time of irradiation	1	12	10/10/41	Immediately relieved, no further attacks	x
23 J.S.	M 49	5 mos.	Constant attacks, cough, dyspnea. bronchiectasis	3	27	7/5/40	No asthmatic attacks, cough, or dyspnea	xx
24 J.A.	M 68	4	30-40 severe attacks; syphilitic heart disease	1	....	1/6/42	....	N.F.

25 A.D.	F 31	2 mos.	Severe constant attack for 2 months	4	12	9/22/41	Completely free of attacks	xx
26 M.A.	F 36	2	20-30 severe attacks monthly	1	....	9/15/40	....	N.F.
27 I.A.	F 56	10	20-30 moderately severe attacks monthly	4	4	6/17/42	Somewhat improved	x
28 F.A.	M 46	39	Severe autumnal attacks	2	....	7/24/40	....	N.F.
29 L.B.	F 33	10	30-40 severe attacks monthly, paroxysmal coughing	10	10	12/1/41	No attacks for 3 months	xx
30 F.B.	M 43	2	Constant attacks plus chronic myocarditis	8	11	11/14/41	Unchanged	0
31 E.C.	F 73	18	Status asthmaticus, 30-50 attacks monthly	5	8	1/20/42	No severe attacks	xx
32 A.C.	F 24	10	8-10 attacks monthly	1	9	1/15/42	Improved	xx
33 H.D.	M 57	14	30-40 severe attacks monthly	3	6	12/19/41	Unchanged	0
34 R.D.	M 5	3	30-40 severe attacks monthly	3	3	6/26/42	Slight attacks	x
35 F.E.	M 61	2	30-40 severe attacks monthly	1	....	2/3/42	....	N.F.
36 F.E.	M 50	3	20-30 severe attacks monthly	3	3	8/4/39	Unchanged	N.F.O.
37 A.E.	F 45	3	30-40 severe attacks monthly	2	2	2/6/40	Somewhat improved	N.F.
38 N.E.	M 30	1	One severe attack in year	1	....	5/5/39	....	N.F.

Continued from page 127

Case No. and Patient's Name	Sex & Age, Years	Duration of Disease Previous to Blood Irradiation, Years	Condition at Time of First Irradiation	No. or Irradia-tions	Mos. Under Treat-ment & Obser-vation	Date of First Irradia-tion	Condition on 10/1/42	Result
39 N.E.	F 21	8	Constant status asthmaticus	4	2	4/15/41	Improved	NFX
40 E.F.	M 44	5	30-40 severe attacks monthly	1	....	3/18/42	....	N.F.
41 N.F.	F 34	8	One severe attack a year	3	8	12/5/41	No attacks	xx
42 C.F.	M 72	0	30-40 severe attacks monthly	1	....	3/4/42	....	N.F.
43 M.F.	F 42	4	30-40 severe attacks monthly	2	1	2/8/39	....	NF
44 J.F.	M 50	34	30-40 severe attacks monthly	3	7	3/9/42	Improved	xx
45 W.F.	M 72	70	Intermittent severe attacks	5	4	12/30/41	Improved	x
46 A.G.	F 36	20	Severe attacks during summer	3	4	2/7/42	Improved	x
47 L.G.	M 56	3	continued status asthmaticus, paroxysmal cough, cyanosis, despite epinephrine and oxygen	2	2	2/27/42	Rapid relief of symptoms following second irradiation	x
48 M.G.	M 32	7	Continued status asthmaticus	5	4	12/3/38	Disappearance of status asthmaticus	NFX

49 A.G.	F 21	6	Severe summer asthma	1	....	....	....	N.F.
50 H.H.	F 31	7	Several light attacks	1	....	8/6/41	....	N.F.
51 J.H.	M 57	8	30-40 severe attacks monthly	7	8	2/7/42	Unchanged	0
52 H.H.	F 65	27	Occasional attacks, seasonal aggravation	1	....	1/7/42	....	N.F.
53 A.H.	F 27	5	30-40 severe attacks monthly	8	8	1/9/42	5-10 mild attacks monthly	x
54 P.H.	F 31	2 mos.	10-15 moderate attacks monthly	7	9	12/10/41	No attacks	xx
55 L.J.	F 34	24	5-10 attacks monthly	6	8	1/28/42	No attacks	xx
56 R.K.	M 21	10	1 or 2 severe attacks monthly	2	24 1/2	9/9/40	No attacks	xxx
57 A.K.	F 15	6	Frequent attacks, epinephrine fast	1	....	3/13/39	....	N.F.
58 H.K.	F 47	22	25-30 severe attacks monthly	11	11	11/11/41	No attacks	xx
59 A.L.	M 6	5 1/2	30-40 severe attacks monthly	2	6	3/24/42	No attacks	NFX
60 A.L.	M 43	3	10-15 mild attacks monthly	3	3	6/27/42	One mild attack monthly	x
61 R.L.	F 7 1/2	7	30 slight attacks monthly	5	10	12/6/41	No attacks	xx
62 E.L.	F 41	10	25-30 severe attacks monthly	8	8	2/6/42	No improvement	0

Continued from page 129

Case No. and Patient's Name	Sex & Age, Years	Duration of Disease Previous to Blood Irradiation, Years	Condition at Time of First Irradiation	No. of Irradiations	Mos. Under Treatment & Observation	Date of First Irradiation	Condition on 10/1/42	Result
63 L.L.	F 22	4	30-40 severe attacks monthly	8	9	12/30/41	5-10 very mild attacks monthly	x
64 J.L.	M 31	3	25-30 moderate attacks monthly	8	10	11/26/41	Occasional attacks and mild wheezing	x
65 M.M.	F 40	4	30-40 severe attacks monthly	6	10	12/19/41	No attacks	xx
66 I.M.	F 49	15	30-40 severe attacks monthly	4	3	7/3/42	5-10 mild attacks	x
67 M.M.	F 35	29	Mild seasonal asthma	1	....	9/9/40	....	N.F.
68 L.M.	M 53	13	10-15 severe attacks monthly	9	9	12/27/41	Occasional mild attack	x
69 D.P.	F 20	3	Apparently moribund, status asthmaticus	0	8	2/7/42	4-5 mild attacks monthly	xx
70 N.P.	M 35	2	30-40 severe attacks monthly	7	10	12/1/41	5-10 mild attacks monthly for 8 months, seasonal; aggravation last 1 1/2 months	x
71 J.R.	F 30	30	50-60 severe attacks monthly	5	10	11/26/41	No attacks	xx
72 V.S.	F 46	11 mos.	20-30 severe attacks monthly	4	7	3/11/42	Unimproved	0
73 E.S.	F 54	11	30-40 severe attacks monthly	2	....	6/29/39	....	N.F.

74 E.S.	M 14	6	30-40 severe attacks monthly	8	8	1/30/42 monthly	2-5 mild attacks	x
75 M.S.	F 34	8	Constant status asthmaticus	4	3	7/13/42	Unimproved	0
76 W.S.	M 60	3	30-40 severe attacks monthly	2	....	5/15/42	....	N.F.
77 J.S.	M 23	10	Seasonal attacks	3	0	4/8/42	No attacks	x
78 S.V.	M 19	14	2-4 attacks monthly, 5-7 days each	8	9	12/19/41	1-2 mild short attacks for 7 months, acute seasonal aggravation for last 2 months	xx
79 C.V.	M 59	5	Spring & fall attacks	2	4	6/16/42	Unimproved	0
80 J.Y.	M 13	13	30-40 severe attacks monthly	10	10	11/28/41	1-2 mild attacks monthly for 8 months, mild seasonal aggravation last 2 months	xx

Summary

Greatly improved	xxx	9
Moderately improved	xx	20
Slightly improved	x	16
Unimproved	0	11
Deteriorated		0
Not followed up—no improvement noted during short observation period	N.F.	19
Not followed up—definite improvement noted during short observation period	N.F.O.	1
	N.F.X.	4

24 not followed up

**Key to Table:

Total   80

Note: The greatly improved group includes only those patients in whom marked improvement has been maintained for more than one year; the moderately improved group, those maintaining definite improvement for more than six months, and the slightly improved group, those showing definite improvement of less than six months duration.

Chapter 11

The Miraculous
Cure of Polio

The effects of photoluminescence on polio may seem to be irrelevant, since polio appears to have been stamped out in North America. Nevertheless, isolated cases still occur, and the ability of photoluminescence to counter such a debilitating disease is simply more proof of its enormous value.

George Miley and Jens Christensen reported on 58 cases of polio. Their results were truly amazing in that, in all the cases treated, they only lost one bulbar case. Bulbar polio (polio of the brain "stem," or lower brain) is 40 percent fatal. The case they lost was near death when seen and was not expected to live. The other six patients with bulbar-spinal polio, also near death when treated, survived.

D.C. was an 11-year-old male who was admitted to the Los Angeles County Hospital, Contagious Disease section, with a diagnosis of progressive respiratory failure secondary to bulbar-spinal poliomyelitis.

The patient was critical, in profound toxemia, with a temperature of 103.2°, and he was extremely lethargic. On 10/18/43, the day of admission, at 9 a.m., he was started on ultraviolet blood irradiation therapy.

The next day, 10/19/43, he was perspiring freely, ate well, and his temperature was down to 100.2°. His condition was definitely improved, but at this time he could not stay out of the respirator.

On 10/20/43, his temperature was 100.8°, and his condition seemed not to be improving further. Because of the lack of continued improvement, the ultraviolet blood irradiation therapy was repeated.

The next two days, 10/21 and 10/22, he showed no improvement and his lethargy continued. It was felt that he was continuing to deteriorate markedly.

On 10/23/43, he seemed to be "possibly slightly improved."

On 10/24/43, it was noted that "toxic symptoms have subsided markedly; very little lethargy if any – definitely improved today."

Two days later, 10/26/43, he was definitely out of danger, though his respiratory paralysis was still present. The next two days he was improving in every way and was getting some respiratory muscle returning to normal function.

By 11/14/43, he was able to be out of his respirator for short periods of time, and his condition was described as "excellent."

Three weeks later, 12/6/43, the patient was out of his respirator and had been out for several weeks, remaining in excellent condition.

There is an interesting historical significance to this case, because it is the first known case of acute bulbar poliomyelitis ever given ultraviolet blood radiation therapy and, we might add, with complete success.

Doctors Miley and Christensen made the following clinical observations concerning ultraviolet polio therapy:

Group One: Acute Poliomyelitis, Bulbar-spinal Type

In three of the seven cases, the swallowing reflex, which had disappeared, reappeared within 24 hours following blood irradiation. In one of these, the swallowing reflex disappeared again 16 days later, but once again returned to normal 24 hours after blood irradiation.

A rapid subsidence of toxic symptoms was noticed in the six cases who recovered.

In two additional cases of this type, consent for ultraviolet blood irradiation therapy was refused, and each patient died.

One of these patients in Group One, who was in a respirator and was in the seventh month of pregnancy, received several blood irradiations and delivered a normal infant at term – the first successful delivery of its kind to occur in California.

Group Two: Spinal Type

This second type of polio is extremely toxic, with rapid progression of muscle weakness and spasm. In this group of dangerously ill poliomyelitis patients, the disease process would normally be expected to develop into severe toxic bulbar poliomyelitis. However, it was noted that as soon as blood irradiation was instituted, the progression of muscle weakness and spasms ceased and in none of the six individuals was there any need for a respirator, although several had beginning signs of respiratory deterioration. In all six patients, a rapid subsidence of toxic symptoms occurred in 48 to 72 hours.

Dr. G.J.P. Barger (Washington, DC) commented on Dr. Miley's cases:

> In my experience in giving 2,500 ultraviolet blood irradiation treatments, I'm glad to be able to confirm Dr. Miley's conservative statements in regard to the efficacy of ultraviolet blood irradiation therapy in the treatment of poliomyelitis of the bulbar type, of which I have treated 23 cases in three years, and six cases of spinal polio. I have compared my eleven bulbar polio cases treated in Washington in 1944 by this method with the series treated by Dr. Miley in 1943 in Los Angeles and find a very close correlation in the results.
>
> When called to give ultraviolet blood irradiation to polio cases, I make a point of prompt response, within one or two hours, for the diagnosis of bulbar polio is often followed by death within a few hours.

> One of Dr. Barger's cases:
> A seven-year-old boy was hospitalized September 12, 1946, temperature 104°, after an illness at home for seven days. He had signs of bulbar involvement on the day of admission to the hospital. The child's physician had told the parents that the boy had one chance in 20 to live.
>
> Ultraviolet blood irradiation of 40 cc of blood was given on the day of admission and, by midnight that day, the temperature had dropped to normal with a corresponding marked drop of pulse and respiration rates and an amelioration of toxic symptoms. The temperature never again rose above 100.4°.
>
> His nurse noted that after the blood irradiation, she had much less need for aspiration of mucus than before. This was most likely due both to a decrease in secretion and a prompt recovery of the ability to swallow noted within 24 to 48 hours in every case of bulbar polio treated.
>
> The child was sent home in two weeks, and 3 1/2 months later, the father reported that "had one not known that the child had had polio, no present sign would suggest it."

The private physician in this case, and in that of another 18-year-old lad treated the previous year for bulbar polio, stated: "These children certainly would not have done as well, in my

Polio Therapy— Summary of Cases*

A. Severe Toxic Type—

Case No.	Patient's Initials	Sex	Age (Years)	Condition at Time of Ultraviolet blood Irradiation	Number of Treatments	Results
1	D.C.	M	11	Progressive respiratory failure, polioencephalitis; patient apparently moribund	3	Improved for 24 hr. after first irradiation, then lapsed into coma; a second irradiation was given 48 hr after first; the lethargic and toxic symptoms began to subside 72 hr. later and had disappeared 96 hr. later; patient has been out of respirator for the last 4 wk.
2	M.N.	F	18	Progressive respiratory failure, polioencephalitis; patient apparently moribund	1	Patient died 62 hr. after the only irradiation
3	C.R.	M	13	Progressive respiratory failure.	2	No toxic symptoms 48 hr. after first irradiation; patient out of respirator part time
4	H.F.	M	17	Patient unable to swallow for two days; nausea, vomiting	1	No toxic symptoms and patient able to swallow in 24 hr.; no further progression of disease process; patient discharged in 14 days in good condition
5	T.H.	F	23	Progressive respiratory failure; patient apparently moribund, 7 mo. pregnant, unable to swallow	2	No toxic symptoms after 24 hr.; patient able to swallow in 24 hr.; in respirator full time; relapse 4 wk. later, with cyanosis, cachexia which disappeared 24 hr after a third blood irradiation; a normal infant delivered at term

6	B.A.	M	16	Patient unable to swallow; respiratory embarrassment, extreme irritability	5	1	No respiratory embarrassment or toxic symptoms 24 hr after irradiation, and patient able to swallow
7	K.C.	F	23	Severe relapse of former respirator patient who had been out of respirator for over two wk.; respiratory embarrassment, anorexia, moderate toxemia, pallor, cachexia present	5	4	No respiratory embarrassment or toxic symptoms present 24 hr after irradiation; no signs of cachexia 5 days later

B. Mild Nontoxic Type—

8	B.W.	M	9	Condition good, but patient unable to swallow for previous 2 wk	6	1	Patient able to swallow within 24 hr
9	D.K.	F	23	Anorexia, mild cachexia, nasal voice tone; patient unable to swallow	22	1	Patient systemically improved and able to swallow within 24 hr.; nasal voice tone unchanged
10	K.K.	F	8	Difficulty in swallowing, nasal voice tone	12	1	Patient able to swallow in 48 hr.; no change in nasal voice tone
11	L.W.	M	4	Patient improving rapidly; spinal symptoms only	14	1	Spontaneous improvement continued

*Miley and Christensen, *Archives of Physical Therapy*, Nov., 1944.

mind, had they not had this form of therapy. They are both complete recoveries today."

Dr. Barger reported:

> The chief pediatrician of this hospital has repeatedly stated to his medical student classes that they had fully expected the death of the first five cases of bulbar polio that they had asked me to treat with ultraviolet blood irradiation, and none of them died.

* * * * *

Dr. Barger also reported an eight-year-old girl desperately ill with bulbar polio, treated by Dr. Miley. During the night following the use of blood irradiation, the child's physician phoned the girl's father to say they did not expect the child to live through the night. The child's father, in gratitude for what actually happened, wrote of her recovery:

> Today, ten months after her nearly fatal bout with a dreaded childhood disease, it is hard to realize that my eight-year-old daughter is almost entirely recovered – walking, skipping rope, vigorous and looking forward enthusiastically to a normal life.

This was his introductory statement in a story written for publication in a widely-read magazine. The story was withdrawn at the request of Dr. Miley, however. In another case, the physician of a six-year-old boy with fulminating bulbar polio wrote to Dr. Miley: "I personally feel that your treatment saved his life."[1]

---

1 *The Review of Gastroenterology*, April 1948.

Chapter 12

Photo-Oxidation
and Cancer

In reviewing the old literature on the use of ultraviolet irradiation of the blood, it was surprising to note that everything from infectious disease to arthritis was treated, but there was no mention of cancer. It seemed odd that these astute medical scientists had not treated cancer patients.

But as this book was about to go to press, a colleague and friend sent me a short pamphlet written by Robert C. Olney, M.D. Olney was a highly-respected surgeon whose articles had been published in the *American Journal of Surgery* (on three separate occasions) and the *Journal of the International College of Surgeons*.

When Olney began to publish his results on photoluminescent treatment of cancer in the mid-60s, he felt certain he would be rejected, so he resorted to publishing a small pamphlet (undated and not copyrighted) to describe his remarkable cures.

Olney was an avid follower of the works of William F. Koch and Otto Warberg, both of whom claimed that a blocking, or impairment, of oxidation enzymes in cells allows fermentation of sugar and that this fermentation in cells is the prime cause of cancer. Olney said that "blocked oxidation" in organisms causes them to become pathogenic (disease-causing) and that this pathological condition is corrected by proper oxygenation; thus, the organisms again become non-pathogenic and, consequently, non-virulent.

Olney stated: "With our present knowledge [1949] it should be possible to prevent and wipe out cancer and serious infectious diseases." That's a bold statement, but Olney backed it up by proving cancer could be completely eliminated with the use of photoluminescence, which he called UBI – Ultraviolet Blood Irradiation.

Olney said:

We are in an era of destructive therapy, powerful poison-ous insecticides, fluoride poisoning, embalmed foods. This is an era of ignoring the principles of healthful living and then attempting to cure everything by taking an array of pills. I believe that the so-called 'accepted' methods of treating cancer are no more successful today than they were 40 years ago. We are entering on an era of prevention and simple, effective treatment of malignant, viral, bacterial, and allergic diseases.

Little did Olney know that 40 years later his studies would still be ignored by the medical establishment.

Following the work of Koch and Warberg, Olney did many studies on blood oxygen levels and proved (again) that oxygen levels in the blood are always low in ill people. Miley had also done studies on oxygen absorption with the use of ultraviolet irradiation of blood in 1939.[1] Olney showed, in every case meas-ured, that blood oxygen levels are drastically reduced in sick individuals. He tested everything from heavy smokers to respira-tory infection to terminal cancer and, without exception, the venous blood oxygen levels were depressed before treatment with photoluminescence and increased after treatment. This effect of ultraviolet light on blood has been known since 1925.[2]

The table on pages 142 and 143 illustrates the vast array of diseased conditions in which blood oxygen levels are depressed. The venous blood oxygen levels before and after treatment with photoluminescence are given.

Doctors Koch and Warberg firmly established that *oxygen deficiency* blocks the basic physiological oxidation processes in the body. This makes the body produce required energy through fermentation instead of oxidation. Koch and Warberg, and many scientists today, agree that this is the pathological basis for malignant, infectious, and possibly even allergic diseases. For instance, hardening of the arteries (atherosclerosis) is not caused by dietary cholesterol but by oxygen deprivation. The theory states that pathogenic organisms become nonpathogenic after oxidation is re-established in them. The remarkable effectiveness of photoluminescence would seem to confirm these ideas. What

---

1 *American Journal of Medical Science*, 197: Page 73, 1939.
2 Harris, Biochemical Society meeting, South Kensington, England, Dec. 7, 1925.

better confirmation than seeing the process work in elimination of disease as the blood oxygen level rises?

In his monograph, Olney reported five cases of cancer out of which he had five recoveries using ultraviolet blood irradiation. One hundred percent is not a bad recovery rate for a disease that is incurable by all "modern" methods.

Case One:

D.P., a 30-year-old white male, was admitted to the hospital with a diagnosis of generalized malignant melanoma (a virulent form of skin cancer).

Eleven years previously, a malignant melanoma had been removed from his right upper arm. When admitted to the hospital by Dr. Olney he had a tumor mass under the skin at the upper left chest just below the clavicle (collar bone). Excision and biopsy revealed that the malignant melanoma had returned. He quickly developed metastases (tumor spread) all over his body and his abdomen became very large from tumor growth. He had difficulty in breathing, had a constant cough and was obviously in serious condition. He was blue in the face and cancer could be felt throughout his abdomen.

The patient was given ultraviolet blood irradiation therapy immediately and approximately every three days for about one week and then weekly. Within three weeks, the large tumor mass in his right armpit disappeared as well as a tumor on the right chest wall; the abdomen became definitely smaller and the tumor masses much less palpable. At the end of six weeks of treatment the patient had no difficulty in breathing; his right leg, which had been extremely swollen, was normal and free of pain; and the abdomen had returned to normal size with no fluid or tumor masses palpable.

Olney remarked about this patient:

This case illustrates the very early, effective results of treatment with the whole program that we are now using for the treatment of all malignancy. This treatment is based upon the evidence that cancer is due to hypoxemia (low blood oxygen concentration) and blocking of oxidation with the development of fermentation of sugars which causes the cancer's growth. It has been shown that all cancers have one common factor and that is fermentation of sugar in the enzymes and cells replacing normal oxidation. Reversal of this process back to normal oxidation is all-important.

Disease	Venous Oxygen Before Photo-R$_x$	After Photo-R$_x$ (24 hours)	After 2nd Photo-R$_x$ (24 hours)
Lymphatic leukemia	48%	65%	
Lung cancer	54%	66.5%	
Uterine fibroid	54.7%	75%	
Heavy smoker	50%	81%	
Severe respiratory infection	48%	65%	81%
Severe diabetic with cerebral vascular accident	48.8%	84%	
Acute mastoiditis and upper respiratory infection	47%	71%	
Congestive heart failure	32%	56%	70%
Asthma with acute myocarditis	48%	58%	68%
Extensive malignancy of the ear and side of head	52%	60%	66%
Severe diverticulosis of the colon, myocarditis and acute respiratory infection	53%	60%	66%
Intestinal obstruction, acute respiratory infection	50%	63%	
Complete acute occlusion of femoral artery at femoral region with impending leg gangrene	38%	73%	

Condition			
Operated on for cholecystectomy and appendectomy	42%	84%	
Coronary insufficiency, respiratory infection	53%	60%	68%
Acute cerebral vascular accident with severe acute otitis media and mastoiditis	40%	56%	66%
Myocarditis, acute respiratory infection, carcinoma of the prostate	34%	48%	60%
Stricture of lower esophagus with almost complete closure of the esophagus and marked dilitation	32%	42%	57%
Cholecystitis and pyelonephritis	47%	71%	86%
Severe neuritis	29%	83%	86%
Thrombophlebitis and cholecystitis	18%	62%	
General carcinomatosis of pelvis from CA of uterus and heavy cobalt treatments	43%	65%	80%

Case Two:

Miss A.F.A., a 64-year-old white female, entered the hospital on April 21, 1953, with a rapidly-developing carcinoma of the left breast. A few years prior, she had cancer of the right breast with a radical mastectomy followed by x-ray therapy. On this second occasion a mastectomy was performed with the removal of lymph nodes in the armpit. It was found that the cancer had penetrated the chest wall into the lung.

Ultraviolet blood irradiation treatments were given at weekly intervals and, after three months, were given at monthly intervals and continued for five years.

"Within a few months there was no evidence of malignancy on x-ray and the patient is now living and well and has had no recurrence of her malignancy," Olney reported.

Case Three:

Mrs. A.R., a 60-year-old white female, entered the hospital on June 22, 1960, with a large mass in the left abdomen which was thought to be an abscess around the kidney. Upon operation it was found that the patient had a large carcinoma of the colon which had ruptured into the abdominal cavity. There were metastatic nodules beyond the primary growth – a serious and terminal cancerous condition.

Following surgery the patient was placed on an intensive program of ultraviolet blood irradiation. She was given four treatments the first week and one each week thereafter for several months. The patient made an excellent recovery and kept up with the blood irradiation treatments, once a month for five years.

Case Four:

Mrs. C.B., a 55-year-old white female, entered the hospital with a diagnosis of tumor of the thyroid gland. During an operation it was determined that she had an adenocarcinoma, so a sub-total thyroidectomy was performed. Following her surgery she was given photo-oxidation treatments, four in the first week and one each week following. The treatment was continued once a month for five years. She remained well and had no recurrence of her cancer.

Case Five:

On April 30, 1969, Mrs. I.W., a 50-year-old white female, entered the hospital for treatment with a large tumor of the uterus that proved to be cancer.

The previous year the patient had been found to have cancer of the cervix and uterus and was given radium and cobalt treatments for a month. Six months later she was re-examined. Her doctor told her that she had a large cancer of the uterus and that nothing could be done. He told her to "just go home and die."

When examined by Dr. Olney she had a large tumor of the uterus and pelvis. She was given photo-oxidation therapy, four treatments the first week and then once a week.

A month later, examination revealed a marked reduction in the size of her tumor and it was felt that she was now well enough to have the tumor removed surgically. A hysterectomy was performed. A pathological examination of the specimen *failed to reveal any viable cancer tissue in the cervix or the uterus*. The patient made an uneventful recovery and did well with no recurrence of her malignancy.

* * * * *

In my own practice, I have also given several cancer patients photoluminescent treatment. Like Dr. Olney, I have found it to be far superior to any other cancer treatment available. However, my treatment has been limited in efficacy by the wavelength of light which I have been using. I have been employing ultraviolet-C, while Dr. Olney used ultraviolet-A. Ultraviolet-C is not the best frequency for treatment of cancer, though, as you will see, it does produce some beneficial results.

At the present time, a new blood irradiation device is being engineered that will use the proper wavelength of light for cancer treatment. The use of such a device will greatly improve the prospects of cancer survival.

Future developments notwithstanding, following are some of the case histories of patients treated in my clinic.

A Friend With Cancer

Patient D.P. is a personal friend, as well as a patient. At about 11:45 p.m. on April Fools Day, 1989, his sister called me, almost hysterical, and stated that she found her brother collapsed in the bathroom, cold, clammy, unconscious, and deadly pale.

The first thing a doctor thinks of in this situation is a massive bleeding episode from something in the intestinal tract. I instructed her to call the ambulance service immediately and have him taken to the hospital, informing them that his doctor's diagnosis is bleeding peptic ulcer with hemorrhagic shock.

The hospital staff agreed with my diagnosis and gave D.P. two units of blood immediately. His hemoglobin was 12 grams and, in my opinion, the blood should not have been given because of the danger of AIDS. Unless the hemoglobin is below eight grams, blood is not warranted. Fortunately, tests done after he left the hospital were all negative for AIDS and AIDS-related diseases. His blood will be checked every three months for at least two years.

Subsequent tests in the hospital, including endoscopic examination of his stomach and CAT scan, revealed a mass which turned out to be a large-cell lymphoma in the top part of his stomach, taking up over one-third of the stomach area. The mass was about the size of a grapefruit.

Against my advice, the patient started chemotherapy 12 days after leaving the hospital. We gave him hydrogen peroxide intravenously three times a week. The peroxide treatment was started before chemotherapy, was continued during, and then also continued after he stopped his chemotherapy. Concurrently, he was given photoluminescent treatment.

He noted that he had absolutely no side effects from the chemotherapy when he was also being treated with photo-oxidation. D.P. said, "When you would leave town, I would always have trouble with the chemotherapy with nausea, vomiting and very severe depression." His doctors, he said, were puzzled that he had so little in the way of side effects from most of the treatment. D.P. said he also felt extremely fatigued and spent a great deal of time in bed when he would take the chemotherapy treatment without having had photo-oxidation.

D.P. lost his hair, as always happens with chemotherapy, and his toenails turned purplish and dropped off. These were the only physical signs of chemotherapy toxicity, other than fatigue, that he noticed during the entire treatment.

Seven weeks after the first CAT scan, another was done, and, much to the amazement of his physicians, the tumor mass had completely resolved. There was absolutely no evidence of cancer being present. Granted the patient was on chemotherapy but I think any qualified doctor would admit that this was a truly remarkable result. D.P. told his doctors that he had been taking peroxide and photoluminescence, and they replied, "Well, perhaps it's a result of both his therapy and ours." (And they may be right. See report on page 149.)

Four-and-a-half months later, in late August or early September, D.P. had a repeat CAT scan, and, again, everything was completely normal with no evidence of any tumor.

D.P. lost twenty-six pounds in the hospital. By October of 1989, he had gained back all of that weight and put on some additional pounds. He feels vigorous and healthy and is now more concerned about keeping his weight down than keeping it up. Parenthetically, it should be noted that his chemotherapy injections cost him over $5,000 per month. The injections cost over $1,000 each. One of them cost almost $2,000. These so-called chemotherapeutic drugs are all listed by the FDA as experimental, yet the patients are charged atrocious fees. If this isn't the biggest swindle in medicine, it certainly has to be close.

Along with his peroxide treatments, D.P. also received photoluminescence on a daily basis. Both therapies should be given for maximum results in treating cancer. A series of cases needs to be done with peroxide alone, photoluminescence alone, and the combination of the two, to determine the relative effectiveness of the two therapies. You will see in the chapter on our AIDS clinic in Africa that the combined therapies are showing quite remarkable results in the treatment of AIDS.

An additional note on patient D.P. He continues to thrive and work full-time, although he is in his seventies, and shows no evidence whatsoever at this time (six years after diagnosis) of ever having had cancer.

* * * * *

The work of Koch, Warberg and Olney proves beyond a doubt that in most, if not all, pathological conditions there is a marked hypoxemia (oxygen lack). This is an extremely important factor in correcting any disease state. Whether the hypoxemia is the cause of the disease or the result of it is not of the utmost importance. The important thing is that blood oxygen saturation must be returned to normal, or above normal, as rapidly as possible if the patient is to be cured.

Hypoxemia or, as Olney called it, "blocked oxidation," followed by fermentation of sugar in cells, is the prime factor in malignant, viral, bacterial, and allergic diseases. We believe, from the reports of the pioneers in ultraviolet irradiation therapy and bio-oxidative therapy, plus our experience, that photo-oxidation will become the treatment of choice for all disease – oxygenation is the key. We also believe that the process will be a factor, in fact the *major* factor, in slowing the aging process.

Will we be able to approach Olney's incredible level of therapeutic success? We think that we can with some modification of our present AIDS treatment program. In the treatment of AIDS, a small volume of blood, only 50 to 100 ml's, is sufficient to obtain a dramatic therapeutic effect. Olney used 500 mls in the treatment of cancer – five to ten times our AIDS treatment dosage. We are currently designing an instrument that will enable us to use volumes of blood comparable to Olney's. We have the added advantage over the pioneers of the '40s of the availability of photo-active compounds that make photo-oxidation ten times more effective than the older methods. Unless the government bureaucracy and the medical-industrial complex unnecessarily complicate the issue, we are facing an enormous breakthrough in the treatment of cancer.

Bone Cancer Secondary to Breast Cancer

M.T. had a radical mastectomy for breast cancer in 1982. In 1986 the cancer returned and spread to the right lung, right rib cage and left shoulder.

In September, 1986, she was given chemotherapy and supposedly had a complete remission. However, in the fall of 1989, even though two weeks earlier she was told that she was completely free of cancer, she developed shortness of breath, weakness, and a collapse of the right lung. It was found that the cancer had spread widely, as described above.

When she arrived at our office, she was in severe pain and appeared to be a terminal case. Her left shoulder and upper arm were so tender that I could not examine them. Due to the spread of the tumor she developed a fracture of the upper arm. The cancer had eaten away the bone in that area.

We started immediate photo-oxidation therapy, warning the family that the prognosis was extremely poor.

After a week's intensive therapy, the patient began to use her left arm, and, after a month of therapy, she could actually extend the arm for us to give her IV's on that side. The pathological fracture of her upper arm healed without any other treatment. The patient's constant cough, secondary to the cancer in her lung, completely cleared after about six weeks of therapy. The chest X-ray showed nothing but scar tissue from all of her previous disease. M.T. also had some metastatic cancer in her right hip necessitating the use of a walker. After two months of therapy, the patient discontinued using the walker.

We are not so sanguine as to think we have cured this sweet and ever-cheerful lady of her far-advanced cancer, but her remarkable improvement speaks for itself.

PIEDMONT HOSPITAL, INC.
ATLANTA, GEORGIA 30309

RADIOLOGY REPORT

ORDERED DATE/TIME 05/26/89 1439	PATIENT NAME					MEDICAL RECORD NUMBER
DATE/TIME TO BE DONE 05/26/89 1439	DOB 01/20/21	AGE 68Y	SEX M	RM. NO / SOURCE ODC	ORDER NO	PATIENT NUMBER
CHECK-IN DATE/TIME 05/26/89 1439	CHECK-IN NO		ATTENDING PHYSICIAN			REPORT RELEASED 05/26/89 1722

ORDERING PHYSICIAN

1938 PEACHTREE RD NW;SUITE
ATLANTA,GA 30309

CLINICAL HISTORY
STOMACH CA

Exam: 74160   CT ABDOMENT WITH CONTRAST

Page :1

A C.T. scan of the abdomen was performed in the routine fashion with intravenous and oral contrast engancement. Comparison is made with the previous study dated 4-7-89. Since the prior exam the large necrotic mass in the fundus of the stomach has resolved and is no longer apparent by C.T. The questionable density seen in the retrogastric region just above the mid body of the pancreas is also no longer seen and there is no lymphadnopathy. No focal masses are identified in the liver or spleen other than a granuloma in the spleen. The renal cysts on the left are again noted. Thre is a small renal cyst in the inferior pole of the right kidney also.

Impression:
Virtual complete resolution of the necrotic mass in the fundus of the stomach and the questionable soft tissue density in the retrogastric region seen on the previous C.T. scan of 4-7-89. No other change is noted. No evidence for lymphadenopathy at this time.

/Read BY/ Eric C. McClees MD
/Released By/ Eric C. McClees MD

MF

Brain Cancer Secondary to Breast Cancer

In February, 1990, J.F., a 53-year-old female was rushed to the local hospital with severe headaches and nausea. The patient was intermittently comatose, and the doctors held out no hope for her survival – even suggesting that if she appeared to be dying, she should not receive any emergency resuscitation. Having been convinced by the doctors that the case was hopeless, the family agreed to "no code."

However, she survived two local hospital stays in spite of metastatic breast cancer to the brain and right lung.

She came to our clinic on April 9, 1990, and was immediately started on photo-oxidation therapy.

About six weeks later, the brain scan was repeated, and there was reported to be "no change." We felt this to be a victory in that we had stopped the advancement of the cancer, which was supposed to have killed her within weeks.

But instead of dying, J.F. began to make a remarkable recovery. Her headaches rapidly became bearable and less frequent; she regained her lost weight; a plantar wart (on the bottom of her right foot) disappeared after having been present for 25 years; two moles disappeared from her left leg. After only one week of therapy, her hair became very shiny and lively; it stopped falling out and began to curl as it always had before she became sick. "Before treatments my hair was dry and looked like a horse's tail. Even the grayness is going away," J.F. reported.

But the most remarkable change in J.F. was the improvement in her vision. She had to throw away her old glasses, because they were "too strong." She needed no correction for distant vision and only a small correction for reading.

J.F. went home for ten days, and, with no therapy, rapidly deteriorated. We were never able to regain the lost ground. Her staggers and headaches had returned, and she died from respiratory arrest.

Would she have remained well if she had not interrupted the treatment? We do not know. No one has ever cured metastatic brain cancer. We cannot claim a successful cure, either, but we are confident that we extended her life.

Chapter 13

OMNI Navigational Systems

Scientists use coherent ultraviolet, infrared, and radio waves to control our satellites and our strategic defenses. However, nature has been doing the same thing for thousands of years – utilizing coherent radiation to control life's processes. The energy obviously must be coherent, or organized, and not just random static energy. It also must be non-linear, meaning that it spreads in every direction and not along a direct line. Ultraviolet light is one type of coherent, non-linear photon energy.

It seems obvious that the world cannot be run by static, or totally disorganized, energy. Coherent waves make for good cybernetic (control) systems, whereas incoherent waves, referred to as "static," do not allow efficient passing of information and therefore lend themselves poorly to cybernetic systems.

It is unlikely that nature is less efficient than man, since both have the same origin and operate under the same laws. So nature uses coherent, non-linear energy waves to run things just as man does to run his satellites.

When corn moths are searching for a corn field, it would seem illogical for God to take time from His busy schedule to direct every moth to its objective. What He has done is to set up a cybernetic radio system in the moth which will direct it to the corn field without further interference, thus saving God's time for more important things, such as supplying new husbands for Elizabeth Taylor's eternal happiness and directing cosmic weather conditions.

If the corn moth has such a wave cybernetic system, then we can assume that other living things have some similar type of electro-magnetic communication system.

Cabbages have, in fact, been shown to emit a pheromone from the large available nomenclature of semiochemicals (defined as any organic molecule that will attract an insect, especially for sex or food). The cabbage looper moth makes his living tracking down this complex signal from cabbages.[1] There is a need to block this self-destructive behavior, this suicidal urge of cabbages, to attract the cabbage looper moth. It is scientifically feasible to do so with our current knowledge of photobiology. This will make for happier cabbages and happier farmers – and the inevitable Save the Looper Moth Foundation.

Like the cabbage, the AIDS virion emits coherent, non-linear photon signals that govern its behavior and so, in order to stop the virus, we need to determine its particular wavelength (71Å) and then block its emission. Photon emission research in living systems is only now beginning, but the evidence is overwhelming that all living systems do transmit and receive coherent signals, just like the cabbage and the cabbage looper moth.

A certified genius and entomologist at the University of Florida, Dr. P. S. Callahan, has postulated the presence of coherent, maser-like radiation in all insects and has, in fact, demonstrated these signals from insect pheromones.[2]

So, although you may not realize it, the lady sitting next to you at a dinner party may be emitting maser-like pheromone radiation which can lead you into the bright light and fire of romantic destruction. However, this maser/photon electro-magnetic conspiracy may also work against her. We now know that her breath emits this non-linear infrared radiation in the 17 to 14 micrometer window when it is modulated between 160 and 260 hertz,[3] which is the same frequency range as the antenna vibrations of mosquitoes and black flies. That is how these nasty little creatures find us when we are out on a pastoral paradise

---

1 The cabbage looper moth can detect one molecule per cubic millimeter of air space. That's one part per 1,000,000,000,000,000,000 – a real nose for food.

2 A maser is a device that produces amplified, coherent microwave radiation of one frequency (or possibly several distinct frequencies) from many frequencies of electromagnetic radiation. Maser stands for microwave amplification by stimulated emission of radiation. Going further, he has even demonstrated maser-like emissions from human breath in the infrared region.

3 Callahan, P.S., Proceedings of the International Conference on Nonlinear Optics, Long, Ireland, May 1988.

trying to enjoy a picnic – we bring them upon ourselves. (Is that the reason women are the first to be bitten?)

Callahan has had rough going with all of this. He can't even convince his colleagues of the unity of these various insect and plant emissions. He remarked, "The problem in the field of entomology is that researchers study insect sounds in one corner of the laboratory and sense in another. For thirty years I've been pointing out that the one modulates the other – to no avail."

If we're going to look at the AIDS virus as an OMNI receiving and/or transmitting station, we need to know a little about the technology. Basically, the AIDS virus is a Hertz transmitter – a wire of a specific length attached to two spheres. This is called a dipole. The spheres at the end of the wire can be moved in and out, shortening or lengthening the antenna, which will change the wavelength of the wave produced. As these spheres reflect the energy back toward the center of the wire, they merge, causing what is known as a "standing wave" – one that appears to be frozen. Hertz, in his original work in 1888, used a wire that was only 2.6 meters long. This length turned out to be almost ideal, not because of the genius of Hertz, but simply because he was poor and, therefore, worked in a very small room – a combination of genius and luck. The original Hertz transmitter was a wire with a sphere attached to each end, as in the figure below.

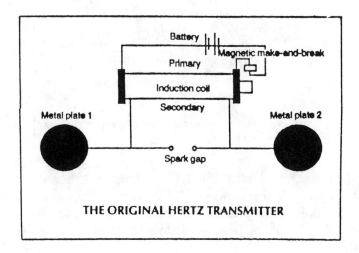

THE ORIGINAL HERTZ TRANSMITTER

Looking at the configuration of the OMNI range apparatus at the Hartsfield International Airport in Atlanta (See illustration on page 157), and comparing it with the AIDS virus (Same illustration), you can easily see that they both have a dipole Hertz configuration. In the OMNI range, the dipole is composed of the bulbs that you see with the ground acting as the other end of the dipole antenna. With the AIDS virus, the large balls (GP-120) are one side of the dipole and the other side is the membrane in which the GP-120 protein is embedded. With this configuration, you have a true dipole, just as in the old Hertz model, with its two bulbs at each end of an antenna. The T4 lymphocyte has rod-like protuberances that make an excellent dipole antenna for receiving messages. We believe, like the corn moth seeking a corn field, this is the way that the brilliantly-designed HIV virus finds its target, the T4 lymphocyte.

T CELL LYMPHOCYTE:

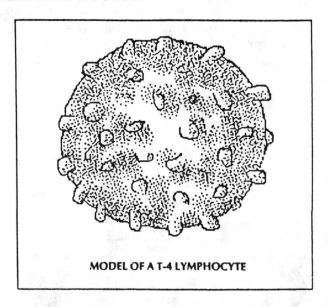

MODEL OF A T-4 LYMPHOCYTE

What we have with the AIDS virus is a beautifully-designed di-electric antenna-OMNI navigational system. It not only looks like the OMNI range at the Atlanta airport, it operates like it. When put in juxtaposition, their similarity becomes obvious:

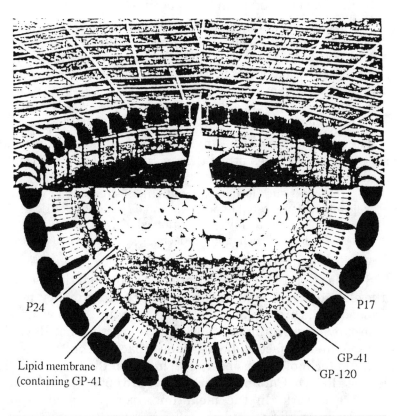

P24

P17

Lipid membrane
(containing GP-41

GP-41

GP-120

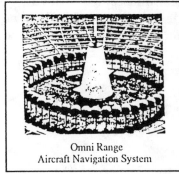

Omni Range
Aircraft Navigation System

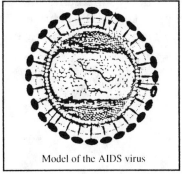

Model of the AIDS virus

Callahan proved, by making a model of the AIDS virus, that it indeed does act as a transmitter/receiver. The GP-120 knobs form one side of the dipole and the GP-41 proteins, embedded in the lipid membrane, form the other. (Or the membrane itself can act as the other half of the dipole.)

Callahan observed from his experiments:

What we can conclude (from these experiments) is that an accurate scale model of the GP-41 – GP-120 AIDS virus shell is of the same wavelength dimensions as the radar wavelength ... [and] does indeed act as a radiation antenna and amplifies the transmitter emission. In other words, given what we know of coherent radiation today, the activity of virion surfaces must be explained by transmittal of frequencies.

There is not likely to be any other explanation for the unique form of virion surfaces other than a frequency explanation. He added:

If there is such [other explanation], this researcher would welcome the suggested hypothesis and be otherwise critical of negative reactions to any explanation which, as is usual, totally ignores the form of such unique appendages in nature.

We now know that the coherent frequency coming from the AIDS virus is 71.42Å. Callahan proposes, through the use of part of the electro-magnetic spectrum, to copy this frequency, put it in the blood and, through harmonics, destroy the AIDS virus without destroying surrounding cells. This is exactly what Royal Rife proposed, and, in fact, did, with cancer back in the 1940s.

Rife's technology has been lost, but with the help of brilliant scientists like Dr. Callahan, perhaps we are not far from imitating Rife's great work. In the meantime, using a combination of bio-oxidative medicine and photoluminescent treatment, we are keeping the AIDS virus and other infections at bay by blocking their signals and by other mechanisms yet to be understood.[4]

---

4 Drawings in this chapter adapted from *21st Technology*, March-April, 1989.

Chapter 14

AIDS:
A Doctor's Story

Most scientists will concede that, in AIDS, the world is facing the worst disease pandemic in its history. Not only is it the greatest biological threat mankind has ever faced, but, because of the uniqueness of the AIDS virus, conventional therapies, such as vaccines or chemotherapeutic agents, are very unlikely to stop the epidemic and the continued devastation of the human race.

The only hopeful field of research, as I see it, is investigation into the cybernetic system of the virus and work toward a coherent wave of visible light, or a wave close to visible light, such as infrared or ultraviolet, to explode the virus through harmonics. A modification of this would be to use the vibrational frequency of the virus and block the virus' transmitter system, thus preventing it from invading T cells, macrophages, Langerhans cells and glial brain cells.

This is technical stuff, however, and may not even be necessary. Photoluminescence appears to do precisely what is needed to stymie the AIDS virus, and thus may be the perfect solution, even though we don't fully understand the problem. Most people don't care what the mechanics of stopping AIDS are (do you care how your car works?), they just care that AIDS be stopped. For them, the following story gets to the bottom line – photoluminescence works on AIDS.

The story concerns a practicing medical doctor who probably contracted the AIDS virus while treating a patient. An associate professor at an American university, the doctor is an honest, competent member of the medical community. His experience with both photoluminescence and accepted methods of treating AIDS is helpful in showing what a difference the former can make – not just in efficacy, but also in reduced cost and frustration. But more than that, his story is about the humanity behind the statistics. It exemplifies why this treatment *must* be made widely available.

* * * * *

I am overwhelmed as I sit here and think about putting my feelings on paper. I wonder if it is even possible. Should I write chronologically what was going on in my mind, to give order to a state of chaos? Or, should I try to write by subject and relive the chaos that I have finally been able to categorize? So much has happened to me in the past year and a half. As for depression, fear, denial, anger – where do I begin?

I am a 31-year-old white male physician who completed residency in July of 1988. As part of an estate-planning recommendation, I applied for life insurance. It was denied on the basis of a positive Elisa for HIV. Further investigation showed my Western blot test was also positive. This was October, 1988.

In retrospect, I had been feeling very poor since February of 1988. My skin test for TB had converted to positive. My physician started me on isoniazid (INH) despite a low-grade elevation in my liver function tests. The INH gave me a raging hepatitis during the spring of 1988. I stopped the INH in April of 1988, but continued to have liver pain. My malaise progressed during the summer to night sweats, fevers, chills, and cramping diarrhea intermittently. I was also experiencing a profound fatigue. I would go to my office daily for a 2 p.m. nap, something unheard of from me during residency training. I could barely walk up two flights of stairs without my legs nearly giving out, I was so weak. My only thought was, "I need a vacation; this job is very stressful."

I started my vacation the day after I had my blood drawn for the insurance physical. I spent October in New England with my wife and two children camping amongst the fall colors. I would crawl into my tent at 7 p.m. and sleep until 9 a.m. My wife only thought that I was overworked. It was upon return from this vacation that I learned the reason for my fatigue.

My initial horror was for my wife, but she tested negative. My children were not tested. My first T4 count was 379, with a T4/T8 ratio of 1.4. My physician felt this was normal and that AZT was not warranted at this time. Yet, I continued to feel terrible.

In the midst of my formative years, my father underwent a major life change. He was a very active veterinarian with a lucrative large animal practice. When I was in the sixth grade, he stopped the practice of veterinary medicine, bought a farm and expected us to help milk cows and raise all organic crops. Needless to say, chemical farming was the norm and my dad was odd man out. Do you know how mean kids can be when they know your dad is a little weird? I hated but loved my father. He would have good years and bad years on the farm, but mostly he had bad years. The bank wouldn't lend him money unless he promised to dump chemicals on his land. One year, the calf crop came in at 32 males and 13 females. Then we had a late spring blizzard that killed all but six of the females.[1] We really didn't do well as a business or as a family. But, I got out by going to college and medical school.

In October, 1988, when I found out I was infected with the Scourge of the Century, I cried. I did not want to die. It wasn't time. I didn't know whom to trust. In my heart, I knew that my father could help me somehow. What a surprising direction he pointed me toward. Over the phone, with big tears in my eyes and my voice faltering, I told him the situation. He calmly asked if I had ever heard of "oxidative therapies." I answered, "no." He told me it may save my life. Then he cried. I had never known my father to cry before. I then found out the reason for the life change of my father. He had been diagnosed as having cancer of the colon at the age of 44. He refused surgery, chemotherapy, and radiation and treated himself with diet, vitamin supplements, and oxygen-enriching therapies. He is now 64.

This began my journey to what I hope and feel is my recovery. I began taking oral H_2O_2 and worked up to a dose of 20 drops of 35 percent food grade three times a day for three weeks. "This is hell," I thought, as I drank each potion. But, I noted a marked purging of my upper respiratory tract and a definite increase in my mental capacity. What a relief to be able to think more clearly.

I pursued the administration of I.V. peroxide and gave myself thrice weekly doses of 250 ml 0.035 percent beginning in January, 1989. Again, I noted an improvement in my overall feeling of well-being. However, my T4 subsets done in February,

---

1 A farmer wants more females than males in calf production. Females produce milk and calves. One bull can service many heifers, or you can skip the bull entirely and use a sperm bank. So a farmer wants females. We think we can increase the production of females by the use of specific light frequencies. It works on fish and chinchillas (see chapter 16).

1989, revealed a drop to 270 with a ratio of 0.3 T4/T8. (The ratio should be 1.0 or higher.) I was frustrated.

I then decided to go to West Germany to be treated with ozone. I was put in contact with a physician in Rosenheim. I arrived in late March 1989, and stayed for three and one-half weeks. I received a total of seven doses of ozone 10,000 gamma via the major autohemotherapy technique. I was also given an I.V. or I.M. drug called Carnivore. This is a naphthaquinone, which is an oxide radical enhancer.

I subjectively felt the ozone was more potent than the hydrogen peroxide. After three treatments, my liver discomfort resolved to the point that I could lie on my right side or my stomach. My energy level improved even more. I was extremely tired immediately after administration of the ozone, an expected side effect. I usually would go back to my room and sleep for two or three hours. However, after my third treatment, I fell asleep and did not awaken for 19 hours. When I climbed out of bed the next morning and realized I had lost 19 hours of my life, I was shocked. I also noticed that my liver did not hurt to nearly the same degree. I also had a tremendous sense of well-being. It was a feeling that I was going to be OK.

It was agonizingly slow to get where I am today, emotionally. I have known fear that left me so cold I would take hot baths and drink hot tea three or four times per day. I occasionally feel this fear, although it is less frequent as I accept my own mortality. I have never known a feeling so lonely and so relentlessly terminal.

When I started oral peroxide, I was in contact with an AIDS patient in Phoenix, Arizona, who was also taking it. He felt that he had been improved and was really upbeat on the subject. I called him about a month later only to find that he wasn't feeling well. A week later he was dead of disseminated coxsackie virus infection. I couldn't get warm – the cold fear had returned.

I was put in contact with a man in Chicago who was getting intravenous hydrogen peroxide in combination with DMSO. His counts had not improved, but they had stopped going down and had remained constant at 300 for the past four years.[2] This wasn't too bad, so I began self-treating with intravenous hydrogen peroxide. My counts fell – my body turned cold again.

---

2 800+ is Normal.

It was this gentleman in Chicago who told me about a video tape that I should see by Dr. Robert Strecker, and he also told me not to trust a thing that comes out of the FDA. I began to investigate further. Dr. Strecker suggested that I call Dr. William Campbell Douglass. At first I was a skeptic. I became less and less trustful of the FDA and the media. But now a new fear. Whom do I trust?

After the ozone treatments in Germany made me feel so much better, I made a decision to pursue any oxidative therapies that were available. I also knew Dr. Douglass was going to Africa to evaluate his photoluminescent instrument. I would wait for his return. Whether or not I would trust his report remained a question.

I pursued "Rife" instruments, glyoxylate, macrobiotics, Interro machines, color therapy, gen elixirs and any crazy notion mentioned as possibly beneficial. In November of 1989, my wife and I separated. We were devastated. My entire free time was devoted to investigating alternative therapies. I had stopped being a husband and a father. My wife was essentially non-supportive of my efforts at self-treatment. She thought it stupid that I wasn't taking AZT, especially in light of the "new" data that AZT was beneficial in the healthy sero-positives with T4 counts between 200 and 500. I had just ordered a new instrument to add to our collection of ozone generators, oxygen cylinders, peroxide bottles, intravenous infusion equipment, syringes, vitamin pills, and macrobiotic cookbooks.

Christmas of 1989 came with our telling her parents we were separated and the reason why. They were shell-shocked. Her mother cried like I have never seen anyone cry. They had always thought we were such a perfect match. I can honestly say that I love my in-laws as if they were my own flesh and blood. I hurt to see them hurt. But they were supportive.

Through much persistence and helpful counseling we are back together again. My wife had been reading some of the literature that I left behind and began asking some hard questions about the FDA's approach to the problem. Also, I have backed off on my pursuit of a "cure" for this plague. I am convinced it is out there somewhere, but right now, I am fine. The photoluminescence has helped tremendously, and I no longer feel the desperation to find something that will work.

I started with a dose of 10 cc heparinized blood exposed to UV light for eight minutes, then reinjected either I.V. or I.M. twice daily. I cannot state in words the difference I feel since beginning this therapy on January 3, 1990. Within an hour after the first injection, my sinuses began to drain profusely. The chronic sinus pain I had lived with for eight years resolved within the week. My energy level went way up. I could go to bed at midnight and wake up at 5 a.m. and feel rested. I had not felt this good since my undergraduate training ten years ago. I also noted that the area over my liver began to itch – yes, itch. It is weird to me, but I firmly believe this to be a sign of healing. Many people who are aware of the situation have commented that my color seems to be better. I no longer have to brush the white film off my tongue every morning.

I am continuing the therapy through February, but have decreased to one dose per day. I am eager, as is Dr. Douglass, to see what my T4 count is at the end of February.

Many people ask how I got the virus in the first place. I personally do not think it should matter. What matters is that I have the virus and to survive I must either eliminate it or control it better than what AZT has to offer. No, I am not homosexual, bisexual, hemophiliac, or an I.V. drug user. Yes, I have stuck my finger with used needles occasionally. I have intubated several AIDS patients in an ICU setting. One of them, I remember, coughed mucous in my face and eyes, despite the goggles I was wearing. Or, perhaps I got the nonrecombinant hepatitis B vaccine in the 70's. It is a feeling among many of my colleagues that HIV is not unlike hepatitis virus. Health care workers are at significant risk. Many have commented that the CDC, NIH, and FDA have their heads in the sand and should get out into the community and see the devastation this virus is having. I concur.

I must end with a discussion of anger. I have a lot of anger, most of it directed toward the major players in the treatment of this disease. I try to keep this anger channelled into constructive activities, but sometimes I cry or lash out. I do not understand how the FDA can say *no* to a phase-I clinical trial of ozone. Their requests for more information are irresponsible in light of the current usage in West Germany.

I do not understand the use of AZT. I step back and take a look at the drug. It was taken off the trial list earlier in the century as being too toxic for cancer patients to endure. Yet, it is rapidly approved for treatment of AIDS. The key word here is rapidly. There has been nothing, and I repeat, *nothing*, approved since. Is that because there is nothing being studied? DDI is no answer. Where is compound Q? Where is CD4? What the hell is going on? Why was AZT moved in so fast?

I do not understand the hoopla behind the drugs to fight the infections. I am grateful for aerosolized pentamidine, acyclovir, etc., but without the virus, we wouldn't need all these new drugs. So, what is going on?

Can you imagine the scenario? AIDS patient enters the system. He is started on 600 mg of AZT per day. The cost is $600 per month. He will probably need two or three transfusions for the anemia – add another $300-$450 per month. Add aerosolized pentamidine at $575 per month. Now, the patient gets CMV. Enter Foscarnet at $20 per day, or $600 per month. We are now spending $2000 per month just to keep the person alive as a walking bag of chemicals. What a boon to the drug makers! But, without the virus, all of the above is unnecessary. Anger.

I keep trying to direct my rage in an appropriate fashion. I am taking care of myself as best I can. I feel sadness for those people with HIV who are not as fortunate as I. When I am more confident in my own survival, perhaps I will break from my current practice and treat AIDS patients with what I didn't learn in medical school. AIDS has become big business, just like cancer. It is amazing how rapidly it has occurred. You can now give your money to AmFAR for AIDS research, but that money is not given to ozone, photoluminescence, Rife, or naturopathic researchers. Why?

On a positive note, my wife has remained seronegative to one year after exposure. I hope and pray that this continues. I also thank pioneers like Dr. Douglass, Dr. Strecker, the International Bio-Oxidative Medicine Foundation, the Medizone Corporation, and any others who are diligently searching for an alternative to the toxic treatment of this and other diseases – and I thought I got an education in medical school!

Well, I must end it here. Sometimes I ramble, and I apologize for that. I must go and "photo-energize." It is time to take care of myself – no doctor in *this* town is going to.

J. B. Will Make Medical History

J. B., aged 35, looked ten years younger than his age. But looks can be deceiving, because J. B. was slowly dying. This highly intelligent, personable, and productive man was dying of AIDS. For two years he had gone along with his doctor's advice concerning treatment until the doctor suggested that he take AZT. J. B. refused, as he felt the drug, from what he had seen among his friends with AIDS, was nothing but a prolonged death sentence. This greatly irritated his doctor, who apparently could not handle any challenge to his authority. Their relationship rapidly deteriorated. J. B. realized that conventional medicine had offered him all that it had, and, because of the doctor's hostile, arrogant, and even impolite attitude toward him, he must make a change in the direction of his treatment. He was referred to us by a colleague from another state, and we informed him, regrettably, that we could not treat AIDS in our clinic because of the great deal of exchange of blood in our type of therapy. We felt it was not fair to our other patients to involve an AIDS patient in this clinical setting. We had been told by the referring physician what a fine young man J. B. was and that it was fervently hoped by the referring doctor and many members of the community that we would do something to help this deserving patient. So, reluctantly, we agreed to see him on consultation, but did not promise him any treatment.

J. B. turned out to be as described, and everyone on our staff felt a need, in fact, an obligation, to help him. A separate area, with completely separate equipment was set up in order to obviate any possibility of cross contamination, and we undertook to help J. B. fight his lonely battle with AIDS. With a P-24 count of 52.3 and a T4 count of 150, we felt that we were probably fighting a losing battle. However, J.B. was willing to do anything in his fight for life, including driving a great distance from another state and spending a considerable amount of time and money at least to attempt to turn his apparently hopeless situation around. His doctor had already told him that his liver function tests indicated he was heading for severe liver disease and that there was nothing else to do, especially if he refused to take AZT. J. B. was always cheerful; he never whined; he never asked for any special attention; he had no bitterness toward the doctor who had treated him so shabbily, and he had a deep, abiding Christian faith that he felt would carry him through. He not only has the support of his church, but also the strong support of his boss in the large company with which he has a very responsible position.

Photo-oxidation was started on April 5, 1990, and his turn-around was nothing short of a miracle. After only seven treatments, his liver function returned to normal, and his HIV indicators all improved dramatically. He improved so remarkably, that after 17 treatments, the therapy was reduced to once weekly. After the frequency of his treatments was reduced, his laboratory values for HIV began varying widely and so twice weekly treatments were resumed. His liver function has remained normal since the first treatment. He remains clinically well and vigorous.

Although the number of AIDS cases in Russia is quite small – the city of St. Petersburg, a metropolis of five million, has only 50 reported cases (.001 percent) – they have treated a few patients with UV light hemotherapy in combination with "anti-HIV virostatic drugs." Their cases did not receive intravenous hydrogen peroxide, other forms of oxygenation of the blood or photosensitive drugs, such as photovite or 8-methoxypsoralen.

Rakhmanova and her co-workers reported on three AIDS cases treated with UV light hemotherapy.

The first case was a 28-year-old male who received five treatments. After two weeks, as in our reported cases, the liver complications, so common in AIDS patients, dramatically improved. Enlarged lymph nodes also receded and the number of blood white cells increased – a very positive sign. Another highly significant laboratory finding in this case was the improvement in the T4:T8 ratio. In healthy people this ratio is 1:1 and is commonly 0.5 or lower. In this patient, the T4:T8 ratio before treatment was 0.43. After treatment, the ratio rose to 0.8 – an excellent improvement.

The second case, "Patient S." was not reported in detail but was said to have had "a very good remission." The third case also was not reported in detail because she only received two treatments and then quit the program. In spite such a short course of treatment, "she had very positive clinical and laboratory changes."

As these patients were also on anti-HIV drugs, the results are inconclusive. It is hoped that they will treat future cases without the use of drugs, now that they have seen these encouraging results from phototherapy.

These cases, plus the case of our doctor-colleague with AIDS, and our experience in Africa, convince us that we have, at least at the present, the treatment of choice for this killer disease.

Liver Function - Patient J.B.

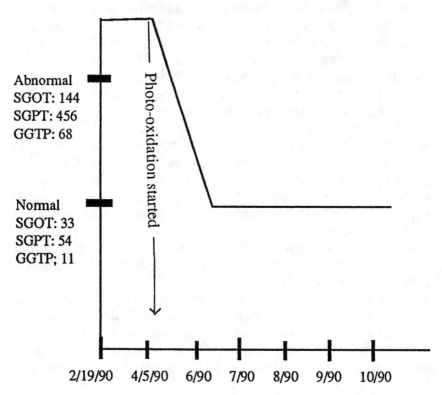

HIV Indicators - J.B.

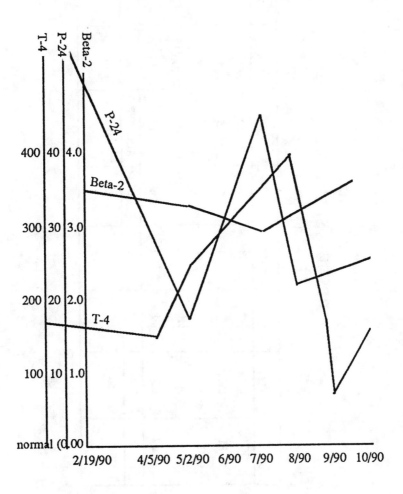

Normal P-24: 0.00
Normal Beta-2: 0.00
Normal T-4: 1000+

RESULTS SUMMARY

J.B. - Lab Results

	2/19	4/5	5/2*	7/26	8/30	10/24	1990
SGOT	144		33	36	50	37	
SGPT	456		54	50	65	56	
GGTP	68		11	15	14	17	
P-24	52.3		18.6	43.9	25.0	27.0	
Beta-2	3.5		3.3	3.1	3.1	3.7	
	FEB	APR	MAY	JUL	AUG (SEPT)	OCT	
T-4	150	128	240	409	190 (90)	184	

Photo-oxidation started (between 4/5 and 5/2*)

*After 7 treatments

Chapter 15

Photo-Oxidation Therapy in Africa

A visitor to a major African city recently had a startling experience when he was taken out to dinner by his host, a physician at the local university hospital. Nearly everyone in the city's best restaurant seemed to know the physician and greeted him enthusiastically – not just the other patrons, who were businessmen, lawyers, and government officials – but also the entire staff of the restaurant. Waiters tripped over each other to make him welcome. Finally, as his guest looked at him in puzzlement at the obsequious display, the doctor leaned over and explained in a confidential tone: "They're all my patients. They all have AIDS."

Just imagine: not only the employees, but all the patrons in the restaurant were carrying the AIDS virus. It reminds me of a story told to me recently by one of my colleagues, who recounted employing three young mountain men in his local community to help him move furniture. After working about half the day and getting fairly well acquainted, the leader of the group said, "Say, Doc, I have herpes and I'm developing a relationship that I think might turn into something. What can you do to help me with my problem?" With that, one of his two friends said, "Yeah, I have the same problem and maybe you can help me, too." The third young man, looking a little uneasy, then replied, "OK, OK – I've got it, too!" The doctor grinned at his three helpers and replied, "Well, gentlemen, I also have herpes, so if you find a remedy, please let me know."

This little story may seem amusing until you think it through. Herpes, a relatively benign, intermittent chronic infection, has an incubation period of probably ten days to two weeks. It is sexually transmitted, like AIDS, but AIDS has an incubation period of up to eight to ten years. How many of the men in this group, 100 percent of whom are carrying a venereal disease, also carry the AIDS virus?

So we must study the situation in Africa very carefully. Is our present herpes epidemic a precursor to an African-style AIDS holocaust?

Health systems in equatorial Africa have become essentially immobilized and panicked by the total chaos and overwhelming burden of the AIDS pandemic: the horror of this deadly disease is rapidly enveloping them; their statistics are unreliable; a good deal of the scientific study is highly suspect; and some countries have taken the road of denial rather than diagnosis in order to protect what little tourism, and consequently income, they have left.

But even those countries that clearly recognize what they're facing are helpless to do much about it. They have few laboratories, hospitals, or sterile supplies, little experience in treating this type of epidemic, and a disastrous lack of medical infrastructure.

After the sad and tragic revelation that the son of the president of Zambia had died of AIDS, the complacency quickly disappeared. Yet, at first, the attempts to analyze and understand AIDS in Africa came across to some people (especially to sensitive African political leaders) as an effort to assign responsibility for an international problem to the poorest continent and to stigmatize Africans for their social behavior. Indeed, some researchers hypothesized at an early stage in the epidemic that AIDS had originated in Africa as a mutation from a monkey virus. This politically-inspired falsehood is still being spread by some U.S. scientists to the unsuspecting American people.

Although it has never been made clear how monkeys were supposed to have given AIDS to humans, one theory implied that monkey bites may have been the cause, or exposure to blood during preparation of monkeys for eating, which is done in the small villages. Several inconsistencies prove this theory totally false. Monkeys are not eaten in the cities where AIDS began. The disease started among the upper classes, who never see monkeys. The villages were not affected by the epidemic in the beginning. Beneath it all was the implication of some bizarre and improper form of sexual contact between Africans and animals. Needless to say, none of this has set well with Africans, who are proud of their race and their nations, just as any other people would be.

Although there was no reason to believe the speculation that Africa was the birthplace of AIDS was based on racism, Africans interpreted it that way. As they saw it, it was an opportunity for the Western world to pass the buck to innocent victims. The revelation that Africa had been contaminated with the AIDS virus through the World Health Organization's smallpox vaccine program only reinforced the feeling that they had been had by the West.[1]

Some have theorized that AIDS is more rampant in Africa because other sexually transmitted diseases, such as genital ulcers and chlamydia, are rampant in African cities. But, it should be noted, these conditions are also epidemic in the United States.

Furious over being made scapegoats, the Africans reacted by denying there was a problem. In late 1988, only seven African nations were willing to submit official figures on AIDS cases to the World Health Organization. A year later, 36 countries were reporting.

Uganda is an excellent country for studying the progress of the AIDS epidemic because the Ugandans (and the Zambians), unlike many other African states, are being honest and open in their discussions of the AIDS pandemic in hopes of getting some help for their people. It is frightening to see the way the epidemic has increased there. In 1983, 17 cases were reported; in 1984, 29 cases; in 1987, there were 1138 cases, and the case load is rapidly increasing. Some speculate that half of all Ugandan adults will have AIDS by the year 2000. But, paradoxically, there may be more hope for Uganda than some of the other countries, because the Ugandans are at least willing to face the fact that they have a terrible problem. Many of the others are not. In fact, it has been reported that anyone who even inquires about the AIDS epidemic in Kenya is liable to be slapped into prison. This type of tragic denial can only make the situation worse.

One gets the true import of the tragedy in Africa when it is realized that some countries only have one doctor for every 25,000 people, and the average expenditure on health is $10 per

---

1 *London Times*, May 11, 1987, p.1.

person per year. Just the test to confirm a single suspected case of HIV infection now costs about $20. Another way of grasping the seriousness of the situation is to compare expenditures in the United States with those in such countries as Zaire. The price of caring for ten AIDS patients in a hospital in the United States, some $450,000, is more than the entire annual budget of a large hospital in Zaire.

Blood transfusions in many African States are used in a fashion similar to what used to be done in the United States back in the early 1900s. Blood may be given for almost any problem, to the point of recklessness, because expensive medications and other treatments simply are not available. Doctors feel they have nothing else to offer. Since blood banks are practically non-existent, relatives of the people who are seriously ill will show up and donate blood on the spot with no testing at all. Consequently, the patient is very, very likely to get some infection from the blood, which can include syphilis, hepatitis, malaria, and, of course, AIDS.

Recognizing the desperateness of the situation and the futility of trying to stop the disease with any semblance of modern medicine, some countries, such as Zambia, have taken the positive step of trying to return their people to the old morality.[2] The Family Life movement of Zambia, with the cooperation of that country's Ministry of Health, distributes a four-page flyer which recommends: "A return to the sexual morality proposed by our best Zambian traditions as well as Christianity, Islam, and Hinduism." In the east African country of Rwanda, the government is putting out almost desperate messages on the radio, pleading with people to be cautious and return to the old ways.

With the establishment of our African AIDS clinic, we are embarking on a new era in medicine. The despondent cry that *nothing works* is no longer true. The advent of bio-oxidative therapy, supplemented with photo-luminescent therapy, means we now have weapons that will enable us to wage an effective holding action against the dreaded viral disease. Although it is not claimed that photo-oxidative medicine is a cure for AIDS, we have seen cases in Africa in the last stages of the disease who

---

2 *Baltimore Sun*, February 5, 1989.

have, after six weeks of treatment, gone back to work and become useful, happy citizens again. Although the comparison is by no means perfect, the best way to conceive of what this combined therapy does is to think of it in terms of insulin for a diabetic. No one claims that insulin cures diabetes, but it enables the diabetic to lead a useful and happy life. Until such time as medicine starts using highly sophisticated electromagnetic and photo-biological medicine, the disease will not be cured. But, as in any war, you have to contain the enemy before you can beat him.

Twenty-two-year-old Amina Nuh died recently. The Kenyan press is not free and information on AIDS is suppressed. The papers merely reported: "She died in the Aga Khan Hospital in Mombasa after a short illness and was buried on the same day in Muslim Cemetery."[3]

Uganda is an absolutely beautiful country, sitting astride the equator. The Ugandans, unlike the Kenyans, enjoy freedom of speech, freedom of religion, and a lively, free, and critical press. The devastation of the civil war which ousted the maniac, Idi Amin, is being rapidly repaired.

The people talk openly about AIDS and its devastation of the population. A photo accompanying an article by Uganda's Director of AIDS Control, Dr. Samuel Okware (*World Health Magazine*), shows a grieving father praying by the graves of his seven children and grandchildren – all victims of AIDS.

Dr. Okware said, "A recent survey of 114 household contacts of 25 AIDS victims showed that only the sexual partners were infected [How, then, are grandchildren catching AIDS?]." He reported that tuberculosis (TB) and other diseases are increasing rapidly, and infant mortality is worsening:

> Socially and economically, AIDS deaths on a large scale among the productive population will threaten agricultural production and development efforts ... political commitment is essential, as is frankness about the disease.

Speaking on education, Dr. Okware said, "The slogan 'zero grazing' caught the public imagination – a folksy metaphor implying that people should not, like cattle, stray from their own

---

3 *Coastweek*, July 28, 1989.

pasture into another." Education is difficult, he noted, in remote communities with little access to television, radio, or newspapers. The president – through his speeches, political organizations, and church groups – is working to educate the people. "Many people find it hard to assimilate the bitter facts about AIDS transmission. We had to soften our campaign with light jokes and comic plays by theater groups."

On condoms, he said: "We have to be cautious about advocating condom use until we fully understand local cultural practices and attitudes."

Dr. Okware concluded on a sad note:

> We are trying to improve palliative terminal care and general maintenance, including psychological and spiritual counseling with the help of church ministers ... admittedly, there is little that can be done for the patients....[4]

We pray (and hope that you will pray with us) that we can, through Peroxide-Photoluminescence therapy, help relieve the incredible suffering we found in Africa.

Whereas the AIDS-infected in the United States die of pneumonia, sarcoma, and common infections such as tuberculosis, the African victim has many other ways to die – such as malaria, chagas disease, yellow fever, and "slim disease" (malnutrition). Unlike many Americans, they suffer in silence, appreciating anything, expecting nothing. Most of the young people between the ages of three and eighteen are orphans, the family remnants of a half-million deaths from the massacres of Obote and Amin. (They both live in opulence in Zambia and Saudi Arabia, respectively.) So life has been very cruel to these young Ugandans. They are kind, gentle people. The injustice of it all could make you cry.

In many African countries, funerals take up a great deal of time. The festivities and ceremonies may take two full days. With the extensive dying from AIDS, you can imagine how much time is expended taking care of the dead, which must be added to the burden of caring for the near-dead. If this condition continues unabated, *there will be no one to grow the food.* The new infrastructure that the Ugandan people have so laboriously and patiently

---

4 *World Health Magazine,* March 1989.

rebuilt – roads, hospitals, hotels, the telephone system, all in less than three years – will be for naught if the AIDS problem is not solved. As a Ugandan friend of mine put it, "Back to square zero." Many other African countries face the same fate.

Uganda was demolished by two homicidal maniacs. The Ugandans picked themselves up only to be cut down again, not by a homicidal maniac, but by a homicidal virus. It must be stopped. Uganda has had enough.

Death is an hourly occurrence in the bush of equatorial Africa. Cheetahs, working in pairs, attack and kill a wildebeest. The vultures then stand by awaiting their opportunity to clean up the remains. A lion attacks an aging hippo, but the hippo manages to escape, only to lie dying, half-submerged, in a pond miles away from the attack. Large hyenas circle for the kill.

And death is also a daily occurrence in the cities. The people are not stalked by lions and cheetahs, but by bacteria, parasites, and viruses. Mosquitoes are ubiquitous in equatorial Africa. The death toll from malaria and yellow fever is awesome. Expensive drugs, such as Chloroquine and Paludrine, are available, but who can afford them? Only treatments that cost pennies are feasible in tropical Africa. Bio-oxidative therapy and photoluminescence offer, for the first time in human history, life and health to millions of people suffering from these devastating diseases.

Although we know the treatment will be effective in a broad range of infectious diseases – the research is there; the results published in the old literature are irrefutable – we are nervous and apprehensive because of the awesome responsibility and the immense amount of confidence that is being placed in us by a few forward-thinking and courageous African doctors. If we are successful, and I am confident that we will be, equal credit must go to these dedicated physicians who have been willing to put their reputations on the line, face embarrassment and even economic and professional harm for entering a new frontier. We hope, with God's help and guidance, to bring about a therapeutic revolution in the third world with these life-giving therapies.

A grandiose and audacious objective? Yes, but we feel it is entirely within the realm of possibility with the weapons we have: intravenous hydrogen peroxide (bio-oxidation) and ultra-violet light (photoluminescence) – together, photo-oxidation, or "photox."

Road to Masaka – Highway of Death

Running south from Kampala to the southern capital of the country, Masaka, is the major artery connecting Uganda with Rwanda and Tanzania. The trucks rumble by incessantly, delivering goods to the heartland of Africa from the major ports of Mombasa, Kenya, and Dar es Salaam, Tanzania.

I remarked to our driver, Sula, how pretty the girls were, dressed in their long flowing dresses with large sashes hanging below the waist, looking very pretty and very African. I asked him if they dressed this way every day, and he said, "Yes, they do," – and giggled slightly. I remarked how wonderful it was that the ladies all dressed so elegantly, even in spite of their poverty, and how proud he must be of the women of Uganda for maintaining their femininity under such grim circumstances. He again laughed nervously.

It dawned on me about an hour later that these were not elegant Ugandan ladies maintaining the country's standard, but were simply truck-stop prostitutes looking for customers. Most of us envision a prostitute as wearing a short skirt that is about two sizes too small and a very tight blouse that bulges in the front. But that is not the Ugandan way. These truckers are known to stop for "tea breaks" two or three times a day, or even more, and to spend their evenings in the same fashion when not driving. This is the way that AIDS has been spread across Africa, having been imported from the Western World through the ports of Mombasa and Dar es Salaam via the smallpox vaccine (see *AIDS – The End of Civilization*). The first cases in Uganda were reported in the sad and suffering town of Masaka. It then spread back toward the other major city, Kampala, where one hundred percent of the prostitutes are now probably infected. All of the prostitutes in Masaka are also infected. So the highway of death flourishes, and the truckers continue to ply their trade and their favorite hobby, which is enjoying the prostitutes, who are also plying *their* trade. In spite of the horrific AIDS epidemic, there seems to be no abatement in their business.

Even more shocking is to see Europeans having dinner with these diseased prostitutes, apparently oblivious – or indifferent – to the danger and the almost certain likelihood that they will be infected by having sex with them. It seems clear that Africa is being blamed unfairly for this epidemic, as it was undoubtedly brought to Africa from Europe by European businessmen, both black and white. Black African businessmen went to Europe, contracted AIDS, and brought it back to their homeland. Although it is a well-kept secret, AIDS started in Africa more than two years *after* it was recognized in the United States.

I visited a Catholic hospital in Masaka and asked the nun in charge of AIDS patients how many cases there were in the area. She replied, "We have no idea." It seems that, through education, even the most backward bush family realizes that there is no cure for AIDS, and so they do not come to the hospital anymore. They don't even come in for testing, because they know the symptoms of AIDS. When they contract AIDS, they die at home, and often will commit suicide, as will the wife or girlfriend soon there-after. In fact, high-tech suicide has come to Africa. The most popular mode of self-destruction is to remove the tiny battery from a digital watch and swallow it; death within twenty minutes – ten minutes if you take two batteries. No one knows, including the pathologists, how many people are taking the "time capsule," as I have dubbed it, because few autopsies are done. Diagnosis is often by supposition and by exclusion. They simply don't have the time, facilities, manpower, or money to conduct autopsies on so many people.

When told by the nun that she didn't know how many AIDS cases there were, we turned to what we felt would be a more reliable source – the man on the street. Our driver, Sula, is close to the people. He told us they were burying between ten and twenty people a day in Masaka. "All you have to do," he said, "is check the graveyards and see how many funerals they're having." As Ugandans do not believe in cremation, this is an accurate way of determining the AIDS death rate (death from other causes is not a significant factor, percentage-wise, due to the overwhelming number of dead from AIDS). I've heard it said that "the AIDS epidemic is disappearing." But it's *not* disappearing – *people* are disappearing.

A young man, whose name is Kaggwa (which means in Luganda: "born by the side of the road"), said he did not have a girlfriend because he was too frightened. "I can't ask a girl if she has AIDS. How can you start a relationship like that?" Young Africans, far more than young Americans, are aware of the danger of AIDS.

We noted a number of casket-making shops along the road to Kampala. Caskets are one of the fastest-selling items in sub-Saharan Africa.

I could tell you much more, but you get the picture. This was the scene in Africa when we arrived to begin our AIDS treatment program there. And though our endeavor was small in size, it was very big in potential. Our hope was to find in photoluminescence the means to eventually arrest the spread of AIDS throughout this vast continent, and to relieve the suffering of millions of people who already had the dread disease.

We stayed for eight weeks and treated many patients. But when we left, we felt we had confirmed the hopes we had for photoluminescent therapy. It had performed well in too many cases to be dismissed as a fluke. It offered very real hope to the millions of suffering and dying in the continent of Africa – and now, I hope, to the rest of the world, also.

I've included several of the case histories of our African patients in Appendix G. By reading them, you'll see a remarkably consistent, beneficial response to photoluminescence in each patient. As usual, no adverse side effects accompanied the treatment, and many lives were prolonged – some indefinitely.

No one can predict the future, but we all like to try. I predict that 20 years from now, and perhaps sooner because of the AIDS epidemic, photo-oxidative medicine will be the mainstay in medicine and replace many of the toxic, useless drugs that are used today. There will always be a place for drugs, but I think almost everyone in the medical profession today admits that they are now overused and abused.

One of the obstacles to this treatment will be its very wide spectrum of therapeutic usefulness. The old adage is, "If it works for everything, it works for nothing." Generally speaking, this is true; but in the case of photo-oxidative medicine, it is *not* true. As you've seen from the case histories, it is indeed a broad spectrum treatment, and there are very few diseases for which it is not worth, at least initially, a try.

With God's help, and the help of our courageous and long-suffering friends in Africa, we will continue to move forward in this exciting, yet terrifying, new era of Medicine Versus Disease.

Section 3

Other Faces
of
Light Therapy

Chapter 16

Ott Man Out

No one in science has made a major breakthrough in science by conducting double-blind studies. Breakthroughs are made by careful observation.
— Dr. John Ott
International Journal of Biosocial Medical Research Vol. 12, 1990

The three great pioneers in the field of photobiology worked about 50 years apart: Neils Finsen, 1890; Emmett Knott, 1940; John Ott, 1980 (and still going strong). Finsen demonstrated the biological properties of ultraviolet light (for which he received the Nobel Prize); Knott discovered the remarkable therapeutic effects of ultraviolet when applied directly to blood; and John Ott brought it all together and demonstrated the awesome effects of various parts of the electromagnetic spectrum, especially the visible spectrum, on all living things, both animal and vegetable.

Of the three, Ott is the only one who has no formal scientific credentials, so his work is not fully recognized by the scientific community; he is the "Ott man out," so to speak. He'll never get the Nobel Prize, but then, neither did Emmett Knott, who certainly deserved it more than some researchers I can think of.

Dr. Ott's monumental work, for which there is "no scientific substance," has shown that the full-spectrum of visible light (and some of the invisible)[1] is absolutely essential for good health. Ott has proven that light is just as important a photosynthesizing mechanism in animals as it is in plants; only the mechanism is different.

At times Ott seemed headed for recognition, even acclaim, by the scientific community. In fact, the heads of the research departments of two large companies confided in Ott that they were very interested, but could not risk any official recognition or participation in his work because it would "subject them to the risk of possible ridicule by other scientists."[2]

Ott's friends, associates and some scientists suggested to him that he just forget about any application of his light research to humans before it drove him into disgrace and became a hardship to his family. One associate made the following sarcastic remark concerning Ott's research: "If John doesn't quit curing cancer by shining a light in everyone's eye, I am not going to be able to accomplish anything for him...."

An article presented to *Survey of Ophthalmology* brought forth the following rejection and comment: "I cannot see that this subject matter belongs in *Survey* at all.... [It] has only the remotest of connections with Ophthalmology."

An application for a small grant from the American Cancer Society elicited the following response: "The advisory committee has recommended disapproval of this application." The reason for this denial was that "results will be difficult or impossible to interpret in any meaningful way (and) no evidence exists or is presented to warrant the belief that such (effective light treatment) exists." And the real killer: "While there is every likelihood that exposure to different kinds of light will affect certain physiological responses in animals, they will only confuse the issue." This is a classic example of scientists not wanting to be confused by new scientific evidence.

You'd think ophthalmologists, of all people, would be up on light science, since the only organ in which they are interested, the eye, is a light receptor. But, strangely, this is not true. As an example, the *Annals of Ophthalmology* carried an interesting and pathetically ignorant editorial in its July 1971, issue entitled "Light: A Double-Edged Sword," in which they stated, "Exposure

---

1 Although humans cannot see ultraviolet, some animals can. The desert iguana and the green sea turtle both have the ability to see ultraviolet light.
2 Ott, *Health and Light*, 1976, The Devin-Adair Co., Old Greenwich, Connecticut, p. 64.

to *normal light* may result in deterioration of the visual cells ... in from seven to ten days' exposure to continuous light of 110 lux[3] intensity *from ordinary light bulbs*." (Emphasis added)

But a study of the article reveals that they are talking about light with a green filter, which, as any competent light physiologist would know, is dangerous to your health. Also, as Dr. Ott points out, incandescent light can hardly be considered "normal light."

Perhaps the most dangerous example of scientific experimentation fueled by ignorance was the study of a treatment devised for retinitis pigmentosa, a hereditary and progressively blinding condition of the eye. It was proudly put forth, in the same editorial mentioned above, that retinitis pigmentosa was going to be treated by putting an opaque contact lens on one eye in order to protect it from the "harmful effects of ordinary visible light" and, thus, double the patient's visual lifetime. The result, however, is literally the proverbial person who's "blind in one eye and can't see out of the other." How effective can a treatment be which treats impending blindness in one eye by covering the other? No one knows, yet, but related research is ominous with regard to the possibilities.

Dr. Irving Geller, the Chairman of the Department of Experimental Pharmacology at the Southwest Foundation for Research and Education in San Antonio, Texas, has shown that rats kept in the dark will develop alcoholism. This clinical condition, induced by keeping healthy light from the eyes, is called "darkness-induced drinking phenomenon."[4] So we can anticipate the possibility of humans treated by covering their good eye becoming blind alcoholics.

In a letter to Dr. Ott from a representative of the American Medical Association in which he was told he would not be speaking for a light symposium (which had been canceled), it was suggested that "ultraviolet radiation does not penetrate the eye" and that the meeting was being canceled "until more scientifically-oriented material is available."

---

3 A lux is one lumen per square meter. A lumen is a unit of light. A lumen per square foot is called a foot-candle, also known as a milliphot. A milliphot is one-thousandth of a phot. You could read this book by a phot's worth of light, but not by a milliphot. Do you see? (Or do you give a phot?)

4 *Science*, July 30, 1971.

Reader's Digest Sheds Little Light

In its June 1990 issue, *Reader's Digest* came out with a scary, in fact near-hysterical, article on the dangers of sunlight caused by its ultraviolet content. The author of the article, psychiatrist David Ruben, may be an expert on sex (*Everything You Ever Wanted to Know About Sex But Were Afraid To Ask*), but he is tragically wrong (as are the American Medical Association and the gaggle of dermatologists who perennially decry the effects of sunlight on exposed skin). He is spreading misinformation about sunlight which will *increase* the incidence of cancer rather than decrease it.

Ruben starts out with the gloomy and truly alarming statistics on skin cancer. He states, "Because of sun exposure, more than six hundred thousand people will be stricken this year and over eight thousand will die; others will damage their eyes." The statistics are correct, but his indictment of the sun is completely unproven and, in fact, incorrect.

Ruben says, "Exposing their bodies to the sun is one of the riskiest things that Americans can do. Medical experts agree that ultraviolet (UV) rays from the sun are the chief cause of skin cancer, which accounts for one-third of all cancer in this country." Medical experts agree on a lot of things, but that does not mean they are correct. Medical experts claim, according to Ruben, the dramatic increase in skin cancer is directly due to "increased voluntary exposure to the sun."

When I was a kid, in Florida, I was exposed six to eight hours of sun per day, and probably close to 12 hours during the summer. My friends and I lived on the beach in nothing but our bathing suits, and none of us, as far as I know, got skin cancer.

It seems strange that man ran around naked for thousands of years without the benefit of sunglasses, clothing, or sunscreen lotions, but the skin cancer problem is singularly a modern one. If God wanted us to wear sunglasses, why were we not born with them? Does your dog or your cat wear sunglasses? Does the hairless Chihuahua have more skin cancer than other dogs? (No, not if they get adequate natural sunlight.)

Dr. Wayne London, a medical researcher at Dartmouth Medical School, agrees with Ott's view of the connection between full-spectrum light and health. On the basis of his research, London has written "Health Benefits of Full Spectrum Light" and "Light and the Immune System," summarizing the connections.

His conclusions include: (1) full-spectrum light benefits rheumatoid arthritis and multiple sclerosis; (2) phototherapy, which uses UV light, is effective treatment for psoriasis; (3) *two types of cancer, breast cancer and intestinal cancer have been associated with a lack of light or a blocking of full-spectrum light*; (4) Patients with candida or chronic yeast infections respond to supplemental full-spectrum light. In addition to clinical improvement, the number of lymphocytes in their peripheral blood also increases.

Ott reported:

> You know, it's common knowledge that when rats or mice are bred in captivity for laboratory experiments, the male has to be removed from the cage before the litter arrives, or he'll cannibalize the young, But we found that when you place the cages in natural daylight or under our full-spectrum lights with radiation shielding, you don't have to remove the male. He'll have a normal parental instinct and help take care of the young. I've taken the same pair of animals and moved them back and forth from natural to artificial light. Under the full-spectrum radiation-shielded lighting, they'll be quite calm and manageable, but you move them back under fluorescence (UV-deficient light) for the next litter and the male will start attacking the young. Their response to the change occurs that fast.

For years we've all been warned of the dangers of full sunlight as a cause of skin cancer and other disorders, and the finger usually points to ultraviolet radiation as the culprit. Yet Ott says we need more ultraviolet – much more than we're getting from artificial light or through windows:

> Without doubt, too much UV is harmful – particularly the short wavelengths, or germicidal ultraviolet, which is mostly filtered out of sunlight by the earth's atmosphere (unless you are a space traveler, you don't have to worry about it – see illustration at end of chapter). But the fear of getting too much ultraviolet is causing people to overprotect themselves from sunlight, to the point that they're creating a deficiency of a very essential life-supporting energy.
>
> This unnecessary fear of ultraviolet is probably a result of general lack of understanding of the different intensities of near, or long-wavelength, ultraviolet and far, short-wave ultraviolet, some of which does reach the earth's surface. Life on this planet exists under a balance of tiny amounts of short-wavelength ultraviolet (comparable to the very low levels of general background radiation) and much higher intensities of

long-wavelength ultraviolet (comparable to that in visible outdoor natural light). It wouldn't take much of an increase in the
short-wavelength ultraviolet to upset that biological balance.
That's why artificial sunlamps (which give off a peak of energy
in the far, short-wavelength UV) become damaging with over
exposure."[5]

Dr. Ott relates a typical example of doctors' ignorance of light
physiology, and consequent maltreatment: "Since the 1950s, it's
been common to treat jaundice in newborn babies with phototherapy using blue light (the wavelength that's most often
missing from our normal indoor lights). Doctors found that the
blue light enabled the babies' systems to break down the excess
bilirubin serum in the blood and correct the jaundice condition.
Westinghouse now makes a very strong blue lamp for use in
hospital nurseries, but it tends to nauseate the nurses! If it's making
the nurses sick, what's it doing to the babies receiving their first direct
exposure to light? Sure, the blue wavelengths help allow the bilirubin in the serum to be absorbed faster, but there may well be side
effects from the lack of other wavelengths found in the full spectrum
of light.

Ott reported:

> Some years ago, I worked with Dr. Jerold Lucey, Professor
> of Pediatrics at the University of Vermont College of Medicine
> and a past president of the American Academy of Pediatrics,
> in some of his photography experiments. He now uses full-
> spectrum fluorescent lights containing the normal amount of
> blue in sunlight. This corrects the jaundice and gives the infants
> a balanced dose of other wavelengths. [And the nurses no longer
> get sick.]

Ott feels that there could be a definite link between some
chronic diseases and a lack of exposure to natural sunlight:

> I'm referring here to what I call biological combustion. My
> studies have indicated that light is a nutrient, similar to all
> the other nutrients we take in through food, and that we need
> the full spectrum range of natural daylight. This is a fact long
> since proven by science. In 1967, a paper presented by
> three Russian scientists to the International Committee

---

on Illumination said, "If human skin is not exposed to solar radiation (direct or scattered) for long periods of time, disturbances will occur in the physiological equilibrium of the human system. The result will be functional disorders of the nervous system and a vitamin D deficiency, a weakening of the body's defenses, and an aggravation of chronic diseases."

That is the condition Ott now calls malillumination, a lack of the necessary amount of sunlight.

Malillumination occurs when wavelengths are missing in various types of artificial light or are filtered from natural light passing through window glass, windshields, eyeglasses (particularly tinted contact lenses or sunglasses), smog, and even suntan lotions. Those particular minerals and chemicals in the individual cells of our bodies that would normally be metabolized by the missing wavelengths remain in the equivalent of darkness.

In other words, Ott believes that energy cannot be extracted from food materials if the proper wavelengths of light are not attained to help break them down chemically.

Dr. Ott has called to the attention of the medical profession the dangers of using tinted eye glasses. He has perfected gray-tinted lenses that will allow the passage of all wavelengths of light. The majority of tinted glasses used today do not permit full transmission of light and are clearly detrimental to the health of the wearer.

Loyola University in Chicago, which had awarded Dr. Ott an Honorary Doctor of Science degree, was interested in accepting Dr. Ott's historic time-lapse equipment and incorporating light research into its department of biology. A site had even been picked out on the campus on which the laboratory would be situated when Dr. Ott learned that the research would be limited to the study of the effect of light on plants and that no experiments on animals would be permitted. Dr. Ott wisely said thanks – but no thanks.

Dr. Ott then offered the laboratory to Michigan State University, but the idea was voted down by the chairmen of *all departments* because "There was no scientific substance to the idea of studying the effects of light on animals...."[6] It's no

6   Ibid, p. 202.

wonder that we call this great pioneer in the field of photobiology the "Ott man out."

* * * * *

To better understand some of Ott's work, take a standard three- or four-cell flashlight and, after turning it on, place it snugly in the palm of your hand. Turn off the lights in the room and turn your hand over to the knuckle side. You will note something that you have undoubtedly noticed before – that the light penetrates your hand; you can see the light through the skin of the top of your hand. From this simple exercise it is obvious that the light penetrates deeper into the tissues than the five millimeters that is cited in standard textbooks of medicine or photobiology. It doesn't take a rocket scientist to figure out that if you can see the light illuminating the top of your hand then, obviously, the light has penetrated your hand.

Keep in mind that is only the *visible* light you are seeing transmitted through your hand. Cosmic rays can even penetrate a concrete wall. We are all exposed, to one degree or another, to all of these rays of the energy spectrum, including the cosmic, x-ray, ultraviolet, the visible light, infrared, and radio. As these energy rays are ubiquitous, and in some cases pervasive, it would be simplistic not to consider that this immense energy ocean that we live in is not related to our vital life processes, both in a positive and negative way.

Dr. John Ott has taken this line of thinking further and compares the skin and some of its receptors with the solar cells used in our satellites to energize them from the sun. Dr. Ott posits that the Langerhans cells are at least one type of these computer-like, semiconductor chips that energize us directly from the sun – just like in plants, only by a different mechanism. The Langerhans cells were thought to be dead squamous cells from the lower levels pushing themselves toward the surface of the skin. But the Langerhans cells, as described by Paul Langerhans in 1868, are obviously not dying or dead squamous cells. He describes the Langerhans cells as being "dendritic cells" located between basal cells and prickle cells. Being dendritic, it's obvious that the Langerhans cells are more akin to nerve cells, or a semiconductor mechanism, than they are to any other type of living cell. A dendrite is, according to *Webster's Dictionary*, "a protoplasmic process in a nerve cell, terminating near the cell body, and which as a rule conducts impulses toward the cell body." The layering of the skin, Ott has observed, is comparable to the layering of different elements used to make capacitors, condensers, transistors, and even simple flashlight batteries. It is very obvious, he noted, that the Langerhans cells fit into this complex electro-photo-magnetic receiving system and are our bodies' biological solar-energy cells.[7] You, just like your pot of geraniums, must have direct sunlight for good health.[8]

Ott's most interesting (and controversial) study concerned the effect of light on the behavior of children.[9] A grade school in Sarasota, Florida, was equipped with Ott's radiation-shielded, full-spectrum fluorescent lights in two windowless classrooms. Two other identical classrooms were equipped with the "standard, cool white" fluorescent lights to be used as controls. Dr. Ott then devised a very clever method of measuring the amount of activity (or hyperactivity) among the children by installing cameras which he could use in time-lapse photography, as he had done for Walt Disney with flowers, to observe the difference in the behavior of the children under the two different light conditions.

Time-lapse photography, in case you're not familiar with it, is a method by which you can see in a matter of seconds an activity that actually transpired over days or weeks. You can therefore watch flowers grow, for instance, by speeding up the film, which has been exposed at regular intervals over an extended period of time.

The cameras recording the activity of the children exposed to full-spectrum light recorded a marked decrease in hyperactivity among the children, with a definite calming effect and an increase in their attention span compared to those exposed to the "cool white," non-full-spectrum lights. In particular, several extremely hyperactive children with "confirmed learning disabilities" calmed down completely and rapidly overcame their "learning disabilities" under the full-spectrum light.

The overall academic achievement level of those exposed to full-spectrum light was superior to those under the deficient lights, and, perhaps the most incredible finding of all, dentists found that those children exposed to full-spectrum light

---

7 *International Journal of Biosocial Research*, Vol. 9, 1987, Part III.
8 Ely, *Dermatology Clinics*, Vol. 7, #1, January 1989.
9 *Academic Therapy*, Vol. 10 (1), 1973.

developed only one-third the number of cavities as those under the "cool white" fluorescent lighting. When the study was completed, Dr. Ott offered to make a donation of lights to the class rooms for the continued health of the children; *the request was refused*.

This bizarre response to an experiment which showed the incredible potential of using light to increase performance (and therefore improve the lives) of school children, is almost impossible to understand unless you understand the bureaucratic mind. Mr. Rick Nations, Director of Research Evaluation and Management Information Services, DREMIS for short (Can you imagine such an important-sounding title for a little county bureaucrat?), said that he couldn't "professionally recommend" Ott's lights and that the matter needed "further study." Well, if the Director of Research Evaluation and Management Information Services of the Sarasota County bureaucracy says he cannot "professionally recommend it," then, obviously, the improved performances never really happened.

So the kids didn't get their lights; their behavior went back to that expected of half-trained animals; and the little bantam bureaucrat went back to his important-sounding job of thwarting any activity even remotely tainted by efficacy.

Dr. Ott's findings were so startling that three representatives of the General Electric Company visited the school and made a careful study of the two different lighting situations and then set out, supposedly, to see if they could duplicate his exciting work. Joining up with two members of the staff of the State University of New York, Philip C. Hughes, of the General Electric Company, did a study and submitted it to the *Journal of Abnormal Child Psychology*.

Their article, entitled "Fluorescent Lighting: A Purported Source of Hyperactive Behavior," states that "the results of this study failed to support the contention that full-spectrum lighting with controls for purported x-rays and low frequency electromagnetic radiation results in a less hyperactive behavior than standard cool white fluorescent lighting."

The study was so sloppy, unscientific, and poorly conducted the editor of the journal, in an unprecedented move, wrote to Dr. Ott to give him a chance to rebut the study. The researchers used a method of radiation shielding different from Ott's. They used seven children in one classroom *with windows*, and with the shades pulled down to within six inches of the bottom of the window, as a control. Ott pointed out that this let in considerable outdoor daylight right at the eye level of the children sitting at their desks. The two types of lighting used in their classrooms were changed back and forth every week, whereas the Sarasota study used very careful controls with no windows or variable types of lighting, so as to have a proper control. In the Ott study, a total of 98 children were involved for five months. Although the GE study ran for two months, there is no way in the world that the study could have ascertained which type of lighting was doing what, when the lighting conditions for a given group of children were constantly switched back and forth every week, meaning that no child remained under the same light for more than five days at a time. Because of the many variables, including lack of a proper control group, any competent high school biology or psychology teacher would give such a study a big fat "F." One General Electric researcher later privately admitted that the whole thing was a conspiracy to discredit Ott's research.[10]

Dr. Ott next approached his local hospital to see if they would be interested in lighting which would reduce their infection rate and, thus, improve the health of their patients. The hospital didn't even reply to his letter.

As Dr. Ott has remarked:

> It would be a big step in the right direction if only they could at least convince the scientific world of the need to include light as a variable in all biological and chemical research studies and not leave this important matter up to the janitor as part of routine building maintenance, as is the present universal practice.[11]

* * * * *

Dr. Ott has done some notable work with cancer patients, and has shown clearly positive results in prolonging life. He and Dr. Jane C. Wright, from the Bellevue Medical Center in New York, advised 15 cancer patients to spend as much time as possible in natural light, without glasses, and to avoid artificial light sources

---

10 *Cutting Edge* newsletter, September 1987.
11 *International Journal of Biosocial Research*, Vol. 7, 1985.

as much as possible, including television. It was the consensus of all those assisting in this program that 14 of the 15 patients had shown no further advancement in tumor development, and several showed positive improvement. One of the 15 patients had not fully understood the instructions and had worn sunglasses. Dr. Ott was met with great antagonism when he presented this small experiment. He was advised that an impending project was going to be called off and that it was not advisable to make any further mention of the previous year's experiment. Even though 14 out of 15 positive responses is not enough, perhaps, to make a closed case, one would think the impressive results obtained would at least elicit interest in further studies – but the only result was extreme hostility.

Dr. Ott also demonstrated the deleterious effects of various colors of light on health. His work with microscopic time-lapse photography revealed that heart cells of chicks, when subjected to red light, weakened and ultimately ruptured. "This again raises the question," he said, "of whether there may be any connection between coronary disorders and the high red content within the spectrum of ordinary incandescent light bulbs." The implications here of the possible effect of ordinary incandescent light on the cardiovascular system are enormous, but no one, as far as we know, has followed up Ott's work in this field.

Many of Ott's experiments showed the effect of light on breeding. Ott found that certain colors, such as pink, have a drastically adverse effect on the development and behavior of animals. In one study on fish, it was found that fish hatched under a pink light were almost all females. Rats that are born under full-spectrum ultraviolet light have a very high survival rate, whereas those under pink fluorescent light only have a 61 percent rate of survival. If mice are left under a pink fluorescent light for six months, their tails will rot off!

The dangers of x-radiation from the ends of ordinary fluorescent tubes and television sets have been demonstrated by Ott. Plants simply won't grow when placed near the ends of ordinary fluorescent lights.

While the so-called experts on light ignore the clear and present danger of x-radiation from the ends of ordinary fluorescent tubes and television sets, they fret about the purported dangers of ultraviolet. It is even claimed, falsely, that ultraviolet light causes cataracts, which, in fact, are caused by the infrared portion of the light spectrum. This has been clearly demonstrated in the case of glass blowers, who get a special type of cataract called "glass blowers cataract." They consistently look into a special furnace, which emits primarily infrared light, not UV light. Most people are constantly irradiated by infrared from incandescent light bulbs, which peak at the infrared portion of the spectrum. The experts ignore this highly significant danger to our health while incorrectly blaming ultraviolet light. Companies vie with each other in their attempt to make their lamps as free of ultraviolet light as possible, while ignoring infrared emissions which can cause not only cataracts but, possibly, heart disease.

Skin cancer, a very common problem in our modern society, is attributed to ultraviolet light by conventional medical authorities. Whereas, it has been shown in Australia, and other areas of very high light intensity, that skin cancer is more common among office workers than it is among those who work in the sunlight.

Ultraviolet deprivation may have caused the dramatic increase in cancer that has occurred in certain areas in Africa. Dr. Albert Schweitzer remarked, when he first started a practice in the Belgian Congo, that the natives did not wear sunglasses and that he rarely saw cancer. After the sunglass craze caught on, because the natives felt it made them appear more modern and sophisticated, cancer became quite common. His nurse remarked that the men pulling their canoes down the river often wore tinted sunglasses and nothing else. Because of their failure to receive sufficient ultraviolet light through their eyes (black-skinned people do not absorb much ultraviolet through their skin), they were not getting this vital energy to their pineal gland and, thus, were "UV light deficient" and susceptible to cancer.

The irrational phobia against ultraviolet light is best illustrated by the action of the very bureau in the government that is responsible for light safety, the Bureau of Radiological Health, Division of the U.S. Food and Drug Administration. All of the fluorescent tubes in the research laboratories, administrative offices, and all other areas, are covered with a special ultraviolet-filtering plastic sleeve, and, for "added protection," there is a plastic diffuser in every fluorescent fixture. So the workers in this bureau, concerned with light and health, get absolutely no trace of ultraviolet light whatsoever and are thereby assured of bad health! This is further confirmation of the Douglass Rule of Inverse Government Action (DRIGA): The more the government tries to solve a problem, the worse the problem gets.

Ott discovered there may very well be a relationship between abnormal fluorescent light, i.e., light that is not full-spectrum, and the onset of leukemia. He reported the case of a school in Niles, Illinois, that had a high rate of leukemia. He visited the school and discovered that the lights they were using, and the curtains filtering the outside light, led to a greenish glow; the fluorescent lights in the room were emitting a strong orange-pink part of the spectrum. After the thin drapes were removed, allowing in full light, and the fluorescent tubes were changed to full-spectrum light, no further leukemia cases were reported.[12]

Ott believes, and has a number of cases to indicate, that the cause of the inability of women to become pregnant is often due to the wearing of tinted glasses. There are 169 "fertility" clinics in the United States that charge as much as $10,000 for therapy and usually fail to help the patient conceive. Dr. Ott relates his experience with six women who were able to conceive after he advised them to throw away their tinted glasses. Estrogen, one of the female hormones essential for conception, has a sharp peak in absorption of ultraviolet light at 290 nanometers. This frequency of light is usually filtered out by sunglasses, and, therefore, the woman's estrogen is not activated properly and she cannot conceive. The same principle may well hold true for the male hormone.

Concerning his landmark research on light and health, Dr. Ott commented:

> The positive results obtained with the various pilot experiments ... produced, in each instance, a reaction of doubt and a suspicion of something not much short of witchcraft.... To suggest that light entering the eyes could have any biological function other than producing vision was like seriously talking – 40 years ago – of man's footprints some day being on the moon.

---

12 Ott, *Health and Light*, op. cit., p. 160.

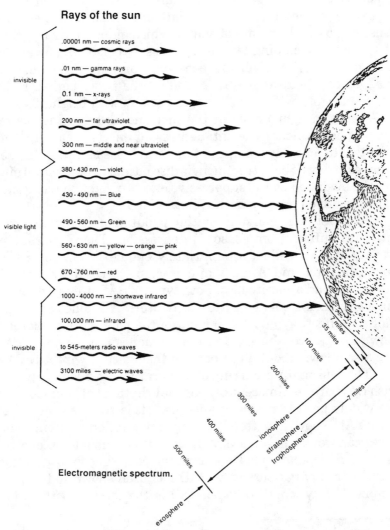

Electromagnetic spectrum.

Reprinted with permission from Sunlight, Zane R. Kine, M.D., <u>World Health Pub.</u>, Box 408, Penryn, Cal. 95663

A SAD Story

Further confirming Ott's work, Seasonal Affective Disorder, or SAD, has become a fairly common, clinically-recognized syndrome which occurs in winter and autumn due to the light deficiency people experience during those times.[13] Patients with SAD are especially sensitive to the short days of winter, which induce in them a cluster of symptoms including fatigue, sadness, hypersomnia (over-sleeping), overeating, carbohydrate-craving, and weight gain. These changes can be reversed by extending the photo period, i.e., the period of light, by five to six hours per day with bright, 2500 lux, full-spectrum light either in the morning and evening hours or evening hours alone. The anti-depressant effects of this treatment take two to four days to appear, and a relapse occurs in the same number of days if light treatment is discontinued. Behavioral effects may be mediated by the suppressive effects of light on melatonin, a light-sensitive hormone in the pineal gland.

At least four percent of the population living at middle latitudes suffers from seasonal depression. There is a summer variety of SAD called, appropriately enough, Summer SAD. It peaks in July and August and is characterized by agitation, insomnia, and appetite loss. Both types of SAD are more common among women from age 21 to 40. Winter SAD is not common in Florida, affecting only 1.4 percent of the population; whereas in New Hampshire, the incidence approaches ten percent. Alaska has also shown about a ten percent incidence of winter SAD.

The seasonally occurring behavioral changes seen in patients with SAD are reminiscent of seasonal rhythms of behavior which are widely prevalent among animals.[14] It is curious that patients with SAD show a pattern of pregnancy different from that of American women in general. Whereas the peak month of conception in the U.S. population is December, the peak time for conception of 219 children of patients with SAD has been shown to be the late summer.[15] Day length (photoperiod) is the most important factor

---

13 *Archives of General Psychiatry*, 1986:43:870-875; *American Journal of Psychiatry*, 143:8, August 1986.

14 Gwinner, *Handbook of Behavioral Neurobiology*, 4:391-408, New York: Plenum Press.

15 Rosenthal, *Annals New York Academy of Science*, undated article.

influencing the timing of fertility in animals.[16] Experimental modifications of photoperiod can drive reproductive cycles independent of the seasons. Thus, extending the photoperiod can produce typical summer behavior in the winter months, and conversely, contracting the photoperiod can produce winter behavior in the summer months.

Armed with this knowledge from animal research, experimenters have tried to extend the photoperiod in the winter with ordinary room light. Such light did not reverse the symptoms of patients with SAD, but they did find that a light intensity of 2500 lux or more was effective in reversing the symptoms, even with an inferior light source.

Although Rosenthal asked the question, "What spectrum or properties of light are necessary or optimal for achieving a therapeutic effect?" He alone of the researchers we've studied has suggested the problem may be ultraviolet deficiency. The literature is replete with discussion of melatonin and pineal gland activity,[17] but, by looking through about 100 references on Seasonal Affective Disorder, I could not find a single reference to ultraviolet light. Although I have not one case to prove it, I feel certain that these patients could be treated with photo-oxidation, once a week for about four weeks, and have their condition completely cleared.

Do you know a patient suffering from SAD? If so, refer them to us and we will be happy to send them to the nearest physician using photo-oxidation therapy. But first tell them to throw away their sunglasses, get full-spectrum light at home and work, and read Dr. Ott's book: *Light, Radiation and You* (1985, Devin-Adair Co., Old Greenwich, Connecticut 06870). It can be ordered from: E.L.C., Suite 101, 3923 Coconut Palm Dr., Tampa, FL 33619.

---

16 Kripke, *Biol. Psychiatry*, 13:335.
17 *Archives of General Psychiatry*, op. cit.

Chapter 17

Other Uses of Light

I always thought that farmers were smarter than doctors, but now I'm beginning to wonder. Perhaps doctors can be excused for overlooking photoluminescence as a therapy because, with the advent of antibiotics, it was generally felt that no more therapeutic technology was needed to solve the problem of infection. But the farmer, who has a tremendous *economic* incentive to use simple, cheap, and safe methods to keep his animals healthy, has also completely ignored the use of light to increase the health of his animals and, thus, greatly increase his profits. For example, a sickly chicken does not produce as many eggs as a healthy one, and the eggs are not as healthy. A sick chicken is underweight and, therefore, as a broiler, will not bring as much profit.

We use chickens as an example of the blindness of farmers to the advantages of proper therapeutic light because it is one of the best examples of ignorance and shortsightedness, drastically reducing profits and causing poor health in farm animals.

We are talking about a major and catastrophic oversight here. For instance, in 1925, Hughes, et al, showed that hens produce *four times* as many eggs when exposed to an ultraviolet lamp.[1] A test at Ohio State University in 1945 showed that if sufficient radiation was used in conjunction with a complete diet, increased egg production and stronger, healthier birds resulted.[2]

Since those early days, highly efficient ultraviolet lights have become available, which have been shown to be extremely

---

1 Hughes, et al, *Poultry Science*, 4:151-156, (1925); Heart, et al, *Journal of Biological Chemistry*, 65, 579-595 (1925).
2 Porter, et al, *General Electric Bulletin*, LD-13.

effective in destroying micro-organisms, mold, viruses, and even protozoa. Lurie found that even airborne tubercular organisms can be destroyed by light radiation.[3]

Extensive tests of the effects of ultraviolet irradiation on chicken egg production were made by Barott and his colleagues.[4] In these experiments, the hens were put in an underground room where absolutely no light was received. The upper air of the room was irradiated with a UV installation comparable to that used in some schoolhouses, except that the intensity of the radiation in the chicken house was two to three times that generally used in schools.

In these experiments, covering a period of three years, the hens in the room with irradiated air averaged more than 17 percent greater egg production than the controls. These results were believed to be primarily from a marked decrease in the number of airborne bacteria and viruses. The incidence of disease was decreased and, as a consequence, the hens were healthier and laid more eggs. As Whitfield said, "sick birds do not lay well," and thus the farmer's profit is reduced. Couch has also reported that "diseased hens are poor layers."[5]

Pullets, i.e., hens less than one year old, usually out-produce older hens. Usually hens lay about 25-35 percent fewer eggs the second year than the pullet year. Barott, et al, showed that hens in an irradiated room produce the same number of eggs the second year as they did the first year. The hens in the unirradiated room produced 11 percent less than they did during their pullet year. So the irradiated hens had a net 11 percent increase in egg production as a result of ultraviolet irradiation.

Carson found that the germicidal radiations were beneficial to young chicks up to eight or ten weeks of age. He stated, "We have rather consistently been able to show an improvement in growth of approximately 30 to 40 grams by use of germicidal lamps. In some instances, this has increased to 60 to 100 grams." Walker confirmed this in finding that chicks exposed to ultraviolet radiation are heavier, more fully feathered and develop much more quickly than those chicks where lamps were not used.

---

3   Lurie, *Journal of Experimental Medicine*, 17; 559-576, (1944).
4   Barott, et al, *Poultry Science*, 30, 409-416 (1951).
5   Whitfield, *Poultry Tribune*, 53,6 (1947).

EGG PRODUCTION OF HENS IN SUBTERRANEAN POULTRY HOUSE[5]

Year	Treatment	Sept.	Oct.	Nov.	Dec.	Jan.	Feb.	Mar.	Apr.	May	June	July	Total Eggs	No. Hens	Eggs/Hen	Percent Increase in Production by U.V. Lamps
1945	U.V. Lamps					635	1007	1120	978	798	192		4730	55	86	19%
	Control					485	1161	767	722	568	136		3839	52	74	
1945–1946	U.V. Lamps	960	2040	1695	1555	1008	1207	1440	1309	1325	160		12699	87	146	16%
1946	Control	800	1781	1627	970	1080	638	840	1043	1114	61		9959	79	126	
1946–1947	U.V. Lamps	154	1677	1976	1825	1776	1373	1211	1251	1164	155		12562	93	135	17%
1947	Control	170	1663	1963	1632	1398	1043	715	1044	1294	219		11141	97	115	

(Egg Production by Months)

5. Barott, H.G., et al, *Poultry Science*, 30, 409-416 (1951).

NUMBER OF EGGS PRODUCED BY BROAD BREASTED BRONZE TURKEYS ON A COMPLETE DIET WHEN SUBJECTED TO ULTRAVIOLET AND INCANDESCENT LIGHT[6]

Month	Incandescent		Ultraviolet No Incandescent		Ultraviolet Incandescent	
	Lot 1	Lot 2	Lot 1	Lot 2	Lot 1	Lot 2
February	36	41	5	9	41	29
March	200	193	205	156	203	211
April	158	141	164	138	192	168
May	178	142	157	151	185	188
June	117	111	111	66	141	159
	689	628	641	520	762	755

6. Carson and Junnila, *Poultry Science 32*, 871-873 (1953).

One of the most impressive reports was that by James Hodges in which he reported on 11,000 chicks exposed to ultraviolet irradiation. He found that the irradiated chicks were 14 percent heavier than an equal number of control chicks. Even more important, the mortality rate was 0.8 percent for birds under lamps and 2.5 percent for the control animals for the same period of time. *Irradiation with bactericidal lamps reduced mortality by 68 percent.* The birds in the irradiated units grew to "range size" in five weeks, compared to eight weeks for those that did not get ultraviolet irradiation. In the chicken industry this can translate into many millions of dollars.

Turkeys, which are even more fragile than chickens, benefit greatly from ultraviolet light. In one experiment, turkey hens were exposed for 14 hours per day for five months to ultraviolet light. There was a ten percent increase in egg production in this group. However, this could not be achieved in turkeys not exposed to natural light at the same time, as they require a long day for proper egg production. Therefore, they must be exposed to the ultraviolet light plus regular sunlight.[6] (See table on pp. 202)

The benefit to turkeys appears not to be as dramatic as for chickens, but more work needs to be done in that field. With chickens, the results are conclusive and the research leaves no doubt of the tremendous economic benefits to be gained from ultraviolet treatment. Veloz found that hens laid 10-15 percent more eggs under ultraviolet conditions, and the incidence of disease was decreased with the poultry being larger, more energetic and better colored.[7] Vaillancourt found a savings of approximately two pounds of feed per bird when the birds were irradiated with ultraviolet. He also showed a considerable decrease in mortality of baby chicks under irradiation.[8]

Even more impressive was a study done by Westinghouse engineers in which they showed that during a period of 19 days of ultraviolet irradiation, 682 hens produced 8,486 eggs; and a

---

6 Carson, *Poultry Science*, 32, 871-873 (1953).
7 Veloz, *Agricultural Engineering*, 26; 67-68, (1945).
8 *Vaillancourt Poultry Supply Dealer*, October 1945.

control group produced only 6,483 eggs – an increase of 28.8 percent![9] Bodwell reported, in a study of 4,000 chicks, that the mortality rate was decreased from about 20 percent to as low as two to three percent by using ultraviolet lamps. This test was done over a period of five years.[10]

Another important advance with ultraviolet treatment is the increased hatch percentages of eggs from irradiated chickens. With current research showing that salmonella actually penetrates the egg shell, this has great import in the chicken industry. The ultraviolet irradiation destroys organisms in the hatching compartments and *on* and *in* the eggs themselves. This is an area crying out for further research because of the serious salmonella problem in eggs in Britain and now beginning to be seen in the United States, especially in the northeast. Sheetz has found that using ultraviolet lamps greatly improves the texture of the eggshell itself. This would imply that the shells would have greater resistance to bacteria such as salmonella.[11]

The increased efficiency of conversion of feed to poultry flesh when ultraviolet light is used is, alone, reason enough for all poultry farmers to employ this method. One test at Preston, Maryland, on 2500 birds over a period of 11½ weeks showed that birds under ultraviolet used 2.97 pounds of feed per pound of meat; those not under the lamps required 3.10 pounds of feed per pound of meat produced. This improved conversion can be explained on the basis that the birds were healthier because of less airborne organisms. However, as the report explained, "probably there are still other beneficial reactions caused by the ultraviolet irradiation for which we do not have an adequate explanation. One of these may be the ionization of the air by the radiation." This report would indicate that perhaps even a negative ion generator could have a positive effect on egg and meat production in the poultry industry. This is another line of research which we intend to investigate.

---

9 *Westinghouse Engineering Note*, ASC-506, August 1, 1950.
10 Bodwell, *Electrical World*, 136, 90, (1951).
11 Sheetz, *Agricultural Engineering 32*, 208-210, (1951).

Chizhevskii reported in 1944 that chickens exposed to negative ions showed an increase in average weight and diminished mortality. Hicks confirmed this in 1956. It is known that ultraviolet lamps will ionize air. Some types of lamps produce more negative ions than others.[12]

The beneficial effects of ultraviolet light on poultry can be summarized as follows:

1. Increased egg production.
2. Increased hatch percentages.
3. A thicker shell and, therefore, a decrease in egg breakage (also a decrease in salmonella in the shells?).
4. Increased growth rate of chicks.
5. Decreased mortality.
6. Pullets feather sooner.
7. Reduced amount of feed per pound of meat.
8. Reduced number and intensity of diseases.
9. Elimination of rickets (vitamin D deficiency).
10. Improvement of the plumage, color of combs, and general appearance of the birds.

To appreciate fully what ultraviolet light can do, not only for the chicken industry but for the entire animal food industry, let's put it in terms of cold cash. As Dr. John Ott has said:

> "Money in the bank" is a ... practical method of measuring research results being used in the poultry industry.... [I]t seems to speak louder than all the papers published in the scientific literature.[13]

Dr. Ott notes that when chickens are kept outdoors, a laying hen is profitably productive for five years. But with the new indoor mechanized methods in poorly lighted buildings, the hens only last for 13 months and then have to be replaced at a cost of $125,000 per 50,000-bird house (1985 dollars).

By extending the laying period of the hens from 13 months to three years by the proper application of UV light, there were the following savings on an annual basis:

---

12 Nagy, *Westinghouse Research Report*, BL-R-6-1079-3023-9.
13 *International Journal of Biosocial Research*, Volume 7, 1985.

	Savings Per 50,000-Bird House
Less feed needed	$ 19,700
8.5 percent more eggs laid	$ 39,800
2 percent fewer eggs cracked	$ 20,000
Larger eggs	$ 7,800
No debeaking necessary (no cannibalism)	$ 4,000
Total	$ 91,300 more profit

When the $125,000 savings in replacement cost is added, there was a total savings of $216,300 – and that's not chicken feed!

There is another interesting advantage to the chicken farmer, due to the current hysteria over cholesterol: The eggs produced under the full-spectrum, radiation-shielded fluorescent lights contained 22 percent less cholesterol than the control eggs.

If left unchecked, one bacterium can produce several billion bacteria in less than 12 hours. Fortunately, the installation of ultraviolet light to stop this massive onslaught of bacteria is relatively simple. In a chicken house, UV light fixtures of a specific frequency and intensity are placed in the ceiling, the distance apart depending on the height of the ceiling. This method of air disinfection is quite efficient because normal air currents in any room cause bacteria to rise and fall from the floor approximately once every minute. No fans are necessary.

As the bacteria are carried into the ultraviolet rays at the ceiling, they are destroyed or weakened. When installed properly and maintained efficiently, this method is equivalent to changing the room air about once every minute. Since the performance of the lamps is affected by many factors, such as other equipment in the room, temperature, draft, dirt, humidity, time, line voltage, etc., the farmer needs to have properly trained and supervised personnel to maintain high production. Remembering that "clean chickens are healthy chickens and healthy chickens are good egg producers," it is worth the farmer's while to have an efficient and knowledgeable company maintaining his equipment. We will be happy to advise anyone interested in increased production of food animals, as well as added safety in the home, the school and other settings where *humans* gather in large numbers.

Landmark studies were done at Duke University almost 30 years ago proving the efficacy of ultraviolet light in the operating room and in any area where people are crowded together.

The Duke study revealed that the colony count of bacteria in operating rooms was reduced to about one to three colonies per plate from an uncontrolled environment which showed 20 to 100 colonies. The authors of the report state:

> Elimination of 75-95 percent of the viable bacteria sedimenting [sic] from the air throughout the room has been followed by a great drop in the infection rate. It is our opinion that most of the bacteria that survived exposure to the radiation were attenuated....

"Attenuation" means that although the bacteria were alive, they were incapable of creating infection.

The Duke studies would apply to any living space where infection is a potential problem. Nursing homes, day-care centers, hospitals, doctors' offices, and theaters are all candidates for this therapy, especially in time of epidemics and as a method to prevent such epidemics.

Ultraviolet irradiation of ambient air, both in the farm setting and with humans, is probably the most under-used, effective technology available today.

Day-Care Centers Need Not Be Disease-Carrying Centers

I have some bad news for you, and some good news. The bad news is that most day-care centers have become centers for the spread of often severe and dangerous childhood diseases. Conditions in them are comparable (to a lesser extent, of course) to the situation in chicken houses — and for some of the same reasons. When you gather many individuals of any species together in a confined space, bacteria and other microorganisms abound. Disease is the almost inevitable result.

The good news, however, is that disease can be prevented in day-care centers, just as it is in chicken houses.

Infants and young children are at high risk for infection under crowded circumstances because they have, even if perfectly healthy, an insufficient immune system. Like an AIDS patient, but certainly not to the same degree, they are truly immuno-deficient. For the first few weeks of life they are protected by "passive" antibodies from the mother, and, as long as they are breast-fed, they are further protected by antibodies in the mother's milk. Once beyond the suckling stage, they are wide open to a large variety of infectious agents abounding in the day-care setting. Infants and toddlers are most at risk because they have the weakest immune systems.

Also, young children are at greater risk for many bacterial infections, especially otitis media (infection of the middle ear), sore throats, tonsillitis, epiglotitis (a life-threatening disease of the vocal cords), bronchitis, and pneumonia. And, because of their weak immunity, they are also at greater risk for skin infections, such as staphylococcus, septicemia (blood poisoning), meningitis, and other blood-borne diseases.

Add to this the not new, but greatly increased, probability of getting serious viral infections, such as hepatitis and AIDS, and it becomes apparent that the situation in day-care centers is very serious indeed. Also, it must be remembered that it is impossible for a day-care center to adequately screen for an infected child, since *before he becomes symptomatic*, he is usually infectious and highly contagious. Studies indicate there is an approximately *twelve-fold* greater risk for hemophilus influenza B infection in children in day-care centers than for children taken care of at home. Meningitis (inflammation of the coverings of the brain by meningococcus) occurs most commonly in children below one year of age and has a high rate of infectivity and mortality. Strep infections, with their sequelae of scarlet fever, nephritis (inflammation and infection of the kidney) and rheumatic fever, which causes serious and life-long heart disease, are potential problems in any day-care center.

One study found that parents of children in day-care centers can expect the infant to be sick nine to ten times a year. A preschool child can be expected to have illness six to seven times a year, usually from respiratory infections.[14]

Infectious forms of diarrhea represent an *eleven-fold* increase in risk in centers with children younger than two years. An intestinal parasite, called giardiasis, has an estimated six-fold increased risk in the day-care setting. *One-third* of hepatitis A outbreaks originate in day-care centers, and over *70 percent of clinical cases can be traced to day-care centers with secondary spread to members of the child's household.* Although hepatitis A is relatively

14 Loda, *Pediatrics*, 49, (1972). Klein, *Review of Infectious Disease*, 8:4 (July-August 1986). Marwick, *Journal of American Medical Association*, 251:10 (March 9, 1984). Trump and Karasic, *Pediatric Annals*, 12:3 (March 1983). Jordan, *Review of Infectious Disease*, 8:4 (July-August 1986).

benign in the young child, the child can become a carrier, and adult members of the family can be very seriously affected.

About 70 percent of day care-related illnesses are respiratory infections. These are normally, in fact usually, spread to the rest of the family by the child.

Infections of the skin, a very common problem in day-care centers, may be spread by way of urine, blood, saliva, and other body fluids. Herpes, furunculosis (boils on the skin), lice, and ringworm are common from the day-care center setting.

One of the more serious infections spread by body fluids is cytomegalovirus (CMV). This viral infection, unknown to most Americans, presents one of our more serious illnesses in this New Age of Pestilence. Although the disease can certainly be serious enough in the child, there is also the risk of the infant who contracts CMV transmitting the infection to mothers who are pregnant or who may become pregnant. This can result in severe birth defects.

CMV can be spread through a respiratory aerosol coughed into the air, ingestion of body fluids through sex or any other close contact, and through blood transfusions. CMV can be found in urine, saliva, blood, semen, breast milk, cervical secretions, and stool.

CMV infection is common in AIDS patients and very frequently leads to blindness. *In view of the fact that 90 percent of AIDS patients are infected with CMV, and 92 percent have evidence of disseminated CMV at autopsy,[15] and also about a 95 percent infection rate with hepatitis B,[16] clearly the AIDS-infected should not work in day-care centers – regardless of your moral and political stand on the emotional issue of quarantine.*

Infants born with CMV disease are afflicted with the most horrible range of deformities one could possibly imagine, including microcephaly (pinhead), retardation, blindness, deafness, quadriplegia, and an IQ at the imbecile or idiot level.

If you don't think that CMV is a clear and present danger to your child in a day-care center, consider the following report on CMV from the *New England Journal of Medicine:*[17]

---

15 *New England Journal of Medicine*, February 2, 1989.
16 Ostrow, et al, *Sexually Transmitted Diseases in Homosexual Men*, Plenum Books, 1983.
17 *New England Journal of Medicine*, May 29, 1986: 314 (22).

Comparing parents of children from ... day-care centers where children have maintained high rates of CMV excretion, as well as parents of children not in day-care centers (controls), for antibody to CMV ... follow-up of seronegative parents revealed that *14 of 67 with children in the day-care centers acquired CMV, as compared with none of 31 controls* (children not in day-care centers). All 14 parents who seroconverted had a child who was shedding CMV in saliva or urine. We conclude that children attending day-care centers often transmit CMV to parents and could be an important source of maternal CMV infection during pregnancy.

Hepatitis B infection is an equally serious problem in day-care centers. The likelihood of becoming chronically infected with hepatitis B virus varies inversely with the age at which the infection occurs, i.e., children are more likely to become "Typhoid Marys" for this disease than adults. Infants infected by their mothers become carriers 90 percent of the time. Between 25 and 50 percent of children infected before five years of age become carriers, whereas only about ten percent of infected adults become carriers. From these statistics it can be seen that a child infected with hepatitis B from the day-care center is highly likely to bring it home and infect parents and other siblings.

The Centers for Disease Control report that infection can occur in settings of "continuous close personal contact, such as in households or among children in institutions." Each year, an estimated 300,000 persons, primarily children and young adults, are infected with hepatitis B virus. An average of 250 die yearly of a fulminant form of the disease. The United States currently contains an estimated pool of *one million* infectious carriers. Approximately 25 percent of carriers develop chronic hepatitis; an estimated 4,000 persons die each year from hepatitis B-related cirrhosis; and more than 800 die from hepatitis B-related liver cancer.

A University of California School of Public Health Study on 255 children in a day-care center revealed that the average child is likely to be ill for *20 to 35 days per year.*[18] That represents the loss of a lot of kid-hours, parent-hours, revenue *from* the schools, and revenue *to* the doctors and drug companies.

---

18 Chang, et al, School of Public Health, University of California, 1978.

What can be done, in the day-care setting, to protect your child from this literal sea of bacterial, viral, fungal, and protozoan organisms? The chicken industry has reluctantly led the way in solving this problem. The solution is relatively simple and cheap and will eliminate probably 90-95 percent of infectious problems in day-care centers. No environment can be made perfect, including the child's own home, but the incidence of infection can be markedly decreased, not only in the day-care center, but in the home, where the child transfers the disease to the parents, by the proper installation of ultraviolet light of a specific frequency and intensity. If the incorrect frequency is installed in an incorrect way, nothing positive will happen. But when properly installed, not only in day-care centers, but schools and other areas where people are closely associated, tens of millions of dollars of infectious disease can be eliminated every year. If the day-care center, or other institution, does not express an interest in this health-saving technique, the next best thing, in fact something you should do anyway, is to install proper ultraviolet lights in your home.

The previously mentioned studies done by the Westinghouse Corporation showed that bacteria and viruses will circulate through a room, rising to the ceiling *once every minute*. Properly installed UV light in the ceiling (occasionally required along the walls) will *very* rapidly eliminate infectious organisms.

Remember, the Westinghouse studies done on chicken coups revealed that proper ultraviolet irradiation caused a *dramatic* increase in egg production. Although you are not interested in your children's egg production, the scientists concluded that the reason the chickens laid more eggs was a decrease in infectious diseases, and a consequent improvement in the health of the egg-laying birds. If it works in a filthy chicken house, it is obvious that it will work in any area occupied by humans.

If you are involved with any institution where people are closely associated with one another, such as nursing homes, rest homes, schools, day-care centers, etc., you can obtain further information on the use of ultraviolet light by contacting:

Health Lighting
P.O. Box 728
Dahlonega, GA 30533-0728
1-800-557-5127

Chapter 18

Phototherapy and
Magnetic Fields

It has long been known that the combination of electromagnetic fields and radiation can stimulate a variety of biological processes. Electromagnetic fields in combination with light will interact with the autonomic nervous system, as well as the endocrine system, and enhance healing. Electromagnetic fields, combined with light, will stimulate physiological processes in the entire body and can be used externally to treat systemic illnesses.

Frederick S. Young and his associates studied the effects of a magnetic field combined with red light phototherapy on AIDS patients. They reported an improvement in the function of the immune system as measured by its ability to fight off various opportunistic infections, an increase in T-cell counts and an increase in the response to antigenic skin tests. AIDS patients, because of their destroyed immune system, lose the ability to respond allergically to skin tests. If the therapy causes them to respond again, this is evidence that the immune system has been strengthened.

The approach of Dr. Young and his colleagues is quite different from photo-oxidation. Whereas we treat blood with ultraviolet irradiation and put it back into the patient, the Young group used only external light. There are many precedents for this approach to the treatment of disease (see chapter 19). As Dr. Young pointed out in his review, bright light can reset the human "internal clock." It can suppress melatonin secretion in humans (melatonin is a light-sensitive hormone secreted from a gland in the brain) and is used in the treatment of seasonal affective disorders. Lengthening the daylight hours can increase the lifetime of hamsters with heart disease by 50 percent, and the regulation of immune processes in the skin is affected by external application of ultraviolet light. Red light phototherapy is now being used at Baylor University to directly attack tumors in human body cavities, such as the bladder and the bronchial tubes of the lungs. Young reports that the proposed basis of his use of red light phototherapy is "a general bio-stimulating effect of visible light on cellular function."

A bio-stimulating effect, similar to that observed with light, has been observed with the use of electromagnetic fields on the body. Young and his associates reasoned that since the reaction of the body is similar with both types of stimuli, perhaps the two combined would be even more effective. We know that electromagnetic field stimulation of the body can alter the nervous system, the endocrine system, the cardiovascular network, the blood, and immune and reproductive systems. Electromagnetic fields have also been shown to enhance a variety of healing and regenerative processes, including limb regeneration in amphibians and spinal nerve cord regeneration in guinea pigs. Electromagnetic field therapy has been used in fracture healing, wound healing, dentistry, and has even been found to reduce the size of tumors in hamsters.

Electromagnetic phenomena obviously play an important role in the regulation of living systems, but, because of the lack of an obvious mechanism for these actions, this research has been ignored until recently. Today, there is much interest in the adverse effects of electromagnetic waves, such as from high voltage electrical lines, video displays, and even, possibly, electromagnetic weapons of war. Young and his colleagues designed their studies to determine whether the effects of whole-body electromagnetic stimulation could be used as a therapy for chronic illness – a healing rather than a destructive process. They used frequencies in the red and radio frequency ranges, but avoided signals in the extremely low frequency (ELF) and microwave regions of the electromagnetic spectrum in order to prevent adverse effects on health.

Their device was used in a study of AIDS to test the hypothesis that low levels of electromagnetism can affect major physiological processes such as T-helper cells.

The apparatus used by Young, et al, consisted of a bed beneath four fluorescent lights with red glass filters.[1] The light intensity was 20 to 100 lux at the body surface, and the red filter was about one yard away from the body of the patient. The base of the bed contained solenoidal electromagnets which produced a magnetic field of 40 Gauss at body level. Coupled to the waist of the patient was an oscillator that produced 400 Khz of electrical stimulation. Patients were exposed to the device a minimum of four hours per day, five days per week.

Routine blood tests, urinalysis, and T cell checks were conducted. The patient's reactivity to skin tests was monitored about every six weeks. Many different antigens were tested which are listed in the second table.

The tables reveal the results of the therapy, which was given independent of any other treatment for malignancy, AIDS, or infection. Six of the eight volunteers had a total of eight active infections of an opportunistic nature when they entered the study. All of these but two cleared completely with the treatment. Skin test reactivity was also dramatically improved. T4 and T8 lymphocyte counts showed improvement in some of the subjects. T cell counts remained improved with retesting over a nine-month period. No subject developed a major opportunistic infection during treatment, which represented a total of over 300 subject-weeks. Routine blood testing and urinalysis remained within normal limits, and the therapy produced no significant side effects after a long duration of treatment. It was noticed that the patients would start to deteriorate two weeks after stopping the photo-magnetic routine.

Dr. Young reported:

> These data suggest that immunologic function was enhanced in the study participants based on clearance of chronic refractory infections which had been unresponsive to standard pharmacological therapy, improvement in skin

---

1 According to Denshah Ghadiali, the world authority on color frequency therapy, blue (440 nm) light, not red, which is at the very opposite end of the electromagnetic spectrum, would be the frequency of choice for a virus infection like AIDS. Yet Young got good results. Maybe the magnetic field changed the quality of the red light, or perhaps the magnetism worked in spite of the presence of the "wrong" color. We have a lot to learn.

Clinical and Laboratory Features of Subjects Receiving Biostimulation

Subject No. & Diagnosis	Symptoms & Signs at Entry	Treatment Toxicity	Clinical Course	Delayed Type* Hypersensitivity — Entry	Delayed Type* Hypersensitivity — 6-8 Weeks	T-helper lymphocytes (mm³) Helper/Suppressor Ratio — Entry	T-helper lymphocytes (mm³) Helper/Suppressor Ratio — Weeks
1. AIDS-post PCPx2	Fatigue Oral candidiasis	None	Energy normal Oral candidiasis resolved off medication	Anergic (mumps	5 positives (T,D,S,Tr,P only)	40/.04	6 Weeks 122/.1
2. AIDS-KS	Fatigue Tinea pedis Onychomycosis KS	None	Energy normal Tinea pedis resolved without medication, Onychomycosis unchanged 6 new KS lesions	2 positives (C,Tr)	5 positives (T,D,C,Tr,P)	150/.3	56 Weeks 304/.46
3. ARC	Fatigue Fevers Night sweats Oral candidiasis	None	Energy normal Fevers, sweats resolved Oral candidiasis resolved off medication	Anergic (mumps	2 positives (D,C) only)	240/.4	7 Weeks 219/.3
4. AIDS-CMV colitis	Severe diarrhea Fatigue Oral candidiasis	None	Diarrhea unchanged Oral candidiasis resolved off medication	Anergic	3 positives (mumps only)	(C,Tr,P) 10/.01	7 Weeks 12/.03
5. AIDS-PCP-KS	Fatigue Pharyngitis KS	None	Pharyngitis resolved Fatigue improved KS unchanged	Anergic	1 positive (P)	20/.1	6 Weeks 0/0
6. ARC	Painful adenopathy Fatigue, Fevers Sweats	None	Adenopathy, fatigue, fevers, sweats all resolved	1 positive (C)	4 positives (T,C,Tr,P)	744/.34	50 Weeks 679/.70

| 7. AIDS-PCP | Fatigue Pharyngitis | None | Fatigue resolved Pharyngitis resolved | Anergic 3 positives (C,Tr,P) Developed KS | Entry 24/.03 | 46 Weeks 92/.30 |
| 8. AIDS-Cyptococcal meningitis-KS | KS | None | Developed additional KS lesions Depressed | Anergic Withdrew at 5 Weeks | Entry 335/.17 | 5 Weeks 497/.30 |

PCP = Pneumocystis Pneumonia; KS = Kaposi's Sarcoma; CMV = Cytomegalovirus
ARC = AIDS-Related Complex

*Merieux Institute Multitest CMI is a 7 antigen plus control skin test battery (T = Tetanus, D = Diphtheria, S = Streptococcus, TB = Tuberculosis, G = Glyceral control, C = Candida, Tr = Trichophyton, P = Proteus) except entry for subjects 1, 3, and 4 when mumps antigen alone used.

Skin Test Reactivity Among AIDS Patients Treated
With Biophysical Stimulation, Untreated AIDS, and Controls

| | Test Antigen[b] | | | | | | | | |
	Control	Tetanus	Diptheria	Strep	TB	Candida	Tricho- phyton	Proteus	Of Control
AIDS patients untreated N=20	0[a]	10	5	15	0	10	0	40	15
AIDS patients receiving biophysical stimulation N=5	0	40	40	20	0	60	80	80	60
Controls N=10	0	90	80	70	60	90	80	50	100

a Expressed as percent positive (# positive/# tested)
b Merieux Multitest CMI used.
c Total positive experimental/total positive control

Adapted from H.C. Lane, with permission.

testing cell-mediated immunity to common antigens, and, in several subjects, marked improvement in the number of T cells, helper cells, and the helper-to-suppressor ratio. Longer courses of treatment revealed that these effects did not appear to diminish with time. Thus, these results indicate a need for further evaluation of the possible use of electromagnetic therapies against AIDS and other chronic illnesses currently refractory to biochemical therapies.

Like other areas of energy medicine, the field of magnetism (and soon, I am sure, light therapy) has been invaded by hucksters who will sell you a black box and promise you a cure for whatever ails you. I called the manufacturer of one of these machines and asked him for scientific references. He replied that "all of the New Age magazines are writing about it."

That wasn't exactly what I had in mind. Not that there is anything wrong with anecdotal reports. Some of the best science gets started that way. But a little clinical observation by trained medical scientists doesn't hurt.

Chapter 19

Chromopathy:
The Power of Color
to Heal

*For centuries scientists have devoted untiring ef-
fort to discover means for the relief or cure of human
ills and the restoration of the normal functions. Yet in
neglected light and color there is a potency far beyond
that of drugs and serums.*
- Kate W. Baldwin, M.D., F.A.C.S.
Atlantic Medical Journal, April 1927

Sir Isaac Newton discovered the scientific significance of the
color spectrum in 1672. He was passing the show window of a
drug store when he noticed light passing through a jar and coming
out the other side as a complete rainbow. He then conceived of
the idea of light being made up of all the various colors, which,
when combined, make white light. Nothing much else happened in the use of light in medicine until 260 years later, when
the brilliant Dr. Denshah Ghadiali worked out his system of
healing with attuned color waves, which he called "Spectro-
Chrome."

Dr. Ghadiali, in his 1200-page masterpiece, *Spectro-Chrome
Metry Encyclopedia,* said:

> Were human beings mere chemical machines, perhaps
> drugs might have had their excuse, but, although these drugs
> act upon the physical part of the human machinery, yet, there
> is something within that machinery which makes it different
> from an automobile. The automobile cannot replace its worn
> parts itself; the human engine does just that in a very effective

manner and thus elevates itself from a conglomeration of dead chemical parts to a marvelous mechanism of self-repairing live factors. The live man and the dead man have a certain difference in their composition and behavior, and it is that which makes man the image of God. There is something within man which is more than the sheer chemical body, and Spectro-Chrome is the only system of healing which, taking into deep consideration the higher life, introduces the remedy for the disorder into those higher vehicles through which man functions as a live entity.

The only scientist between Newton and Ghadiali that we need mention with respect to the biological effects of the light spectrum is Edwin Dwight Babbitt, M.D., L.L.D., who did his research during the late nineteenth century.

In his book, *The Principles of Light and Color*, 1878, Babbitt wrote:

It is quite time that the wonder of light and color, which is invisible to the ordinary eye and which is capable of being demonstrated by spectrum analysis and otherwise, should be made known, especially as so many mysteries of nature and human life are cleared up thereby and such marvelous powers of vital and mental control are revealed.

Babbitt was ridiculed and laughed at by his colleagues, and the *Scientific American* wrote sarcastic articles about his work. Yet, he was clearly years ahead of his time. He wrote on chromo-chemistry and chromo-therapeutics and the effects of various types of light on vegetables. Because of similar work by Pleasonton, who used blue glass to increase the output of grapes, the whole business was called by the scientific community the "Blue Glass Craze." Babbitt died in poverty – a broken-hearted man.

Denshah Ghadiali was born at about the time that Dr. Edwin Dwight Babbitt was being pilloried by the scientific community. About 1890, following a very dramatic vision, Ghadiali set out on his work on the use of color in the healing art.

It did not take Ghadiali long after his medical training to become disillusioned with "medical science." "How does the pill know where to go?" he asked.

He began delivering lectures against ingestion of chemicals and drugs, speaking in rather harsh terms against the shortcomings of so-called modern medicine. At one of these lectures a doctor came up to him after his talk and said:

It was a beautiful lecture you gave; you are a wonderful young man, but say, though I admit we are a bunch of boobs as you aver, it would be better if you kept your mouth shut 'til you had something more effective in medicine to offer!

Ghadiali took this admonition to heart and said nothing more until he had developed "something more effective." His great therapeutic breakthrough came at an unspecified date with the treatment of an old friend who had great admiration and respect for his medical abilities.

One evening he was called to the home of this friend, a 29-year-old woman. He found her to be severely ill with mucous colitis, a severe form of diarrhea. She was in severe pain and losing a great deal of blood from this condition. "Jerbanoo was having in a day a hundred times more or less such urge and life was speedily ebbing," he observed.

Although she had confidence in the 24-year-old Dr. Ghadiali, she insisted on staying with the prescriptions given to her by the learned physicians of the community. She was taking a conglomeration of opium, catechu, chalk, bismuth subnitrate, and chloroform. But after three days in a continuing decline, she lifted her hands to the young physician and said, "Oh Denshah, save me!"

Ghadiali said:

> Medically, she was beyond recovery, and I said with a sigh, "Call on the Almighty to save you, dear girl, I have no power; no medicine of which I know can be of service to you, but, if you let me, I shall endeavor to do the best otherwise." She nodded her consent, and promptly I threw out the brandy bottle and the drug mixture.
>
> Here was my opportunity to test the chromopathy of Edwin D. Babbitt. The woman was dying – she was anyhow as good as dead. I could not kill her further if I failed.

Ghadiali went to the market and purchased a kerosene oil hurricane lamp similar to what, in those days, was used by the Highway Department doing night work, and some colored pickle bottles to act as color slides.

Applying to her abdomen colors that he had surmised would be appropriate in this type of disease, within twenty-four hours her bowel movements began to be reduced, and within three days she was out of bed.

Ghadiali said:

I had enough to wash my hands of the British Pharmaco-poeia. Drugs were not precise, they were of no use to me; here was my experience, my proof. Other experiences followed this one, and my mind whirled with the intoxication of using the higher forces of the physical plane for the alleviation of the ailments of humanity.

No medical doctor can afford to be so ignorant as to deny the powers of light. What they say, however, is that the potency of light is solely in the infra-red, and the ultraviolet rays and the visible spectrum, where color is produced, have no value in therapeutics. How and where they concede such an errone-ous idea is a mystery. The beauties of colors cannot be useless, unless the scientists believe their purpose is solely aesthetic, and they were made by the Creator for vainglory and decora-tion.

Ghadiali relates the story of a 19-year-old boy, who was brought to his institute in New Jersey, almost moribund from advanced pulmonary tuberculosis. Although it appeared to be a hopeless case, Dr. Ghadiali agreed to keep the boy overnight and treat him with color therapy to his lungs.

He placed him under the proper color frequency at one o'clock in the afternoon, and by 2:30 that afternoon, he was sitting up in bed; by 5:30 he was taking a full meal. The next morning he walked downstairs with the help of his mother and commenced to eat his breakfast.

His father, on returning that morning, was astounded to see his son quietly eating, with no apparent discomfort. He said, "Colonel, I now see why they call you a magician; I never thought I might see Paul alive again."

Thus was started a remarkable career by a remarkable man, who deserved a Nobel Prize for his medical advances, but received his reward only in the afterlife. He died in 1966 with most of the world never having heard of him or his remarkable therapy.

The idea of shining a specific color on a diseased part of the body for treatment will sound preposterous to most, but think about it: ultraviolet and infrared, both invisible parts of the electromagnetic spectrum, are readily recognized as useful in medicine, so why should anyone be surprised that the rest of the spectrum, i.e., visible light, is also useful? Applications of light, the color red, for instance, which has a frequency of 436,803,079,680,000 oscillations per second, is just as much a form of energy as the note high C, which has an oscillation frequency of 4,096 oscillations per second. X-rays, gamma-rays, delta-rays and magnetism are all forms of oscillatory energy. As far as we know, magnetism has the highest oscillatory rate which is 18,446,744,073,709,551,616 oscillations per second.[1] It is hard for the mind to grasp how anything can take place that many times in only one second. That's a long way from the oscillation of the lowest inaudible (to humans) sound which is a mere two oscillations per second.

I am dragging you through all of these numbers to help you understand colors which are not what you *see*, but what they *are*: vibratory energy. The average physician, unfamiliar with color therapy, would say: "That can't possibly have any effect – it's only color – it doesn't penetrate the skin." But what the average doctor doesn't realize is that the body, like all living things, has an electromagnetic aura around it, and light therapy works very powerfully on the aura. It does not have to penetrate the skin to have an effect.

Before you decide this is some sort of New Age, satanic and/or goofy quackology, let's relate a most incredible medical case in order to put your feet firmly upon clinical grounds. Then we will come back for a closer look at the science of chromopathy.

Kate W. Baldwin was a distinguished physician from Philadelphia where she was graduated from medical college in 1890. She was the Senior Surgeon at Woman's Hospital of Philadelphia for twenty-three years. She was a member of the American Academy of Ophthalmology and a member of the American College of Surgeons.

Dr. Baldwin was called to testify on behalf of Denshah Ghadiali, M. D., the originator of color therapy. Dr. Ghadiali was being prosecuted (persecuted) for using color therapy on his patients. He had been jailed and as much of his equipment as could be found was smashed with sledge hammers by government agents. Persecution of doctors who dare to be different obviously is not new. This trial took place about 1930.

Dr. Ghadiali, who was defending himself, asked his physician-witness about a certain burn case, a little girl by the name

---

1 That's 18 quintillion, 446 quadrillion, 744 trillion, 73 billion, 709 million, 551 thousand, 6 hundred, and 16 oscillations per second.

of Grace Shirlow. Dr. Baldwin produced pictures showing the gruesome result of the burns of the clothing and skin, which covered four-fifths of the torso of the child. Dr. Baldwin said, "It was not only the skin that was destroyed, but the fascia of the muscle." Dr. Baldwin went on to say that the skin was completely gone and the coverings of the muscles were exposed over a great deal of the body.

Dr. Ghadiali asked: "By looking at that case, from your surgical experience, did you believe that any method known to medicine and surgery, could keep that child alive?"

Dr. Baldwin replied:

> It is generally conceded, with that much of the body burned, the patient cannot be saved. This was a hopeless, fatal case.
>
> In fact, I got there about twenty-four hours after the surgeon had been called in. He went out very legitimately, just simply wrapped it up in gauze and cotton to protect it, as he was quite justified in saying, "there is no use in trying to do anything with this!" In fact, the dressing was so tightly pressed into the raw surface that it was two weeks before I succeeded in getting it all off, as I would not force it off. I had to wait until the healing process took place underneath and it loosened up, because if you pulled off the dressings, you would pull off new tissue as well as old.

The indictment claimed that Denshah Ghadiali "did feloniously steal $175 from Housman Hughes by falsely representing and pretending that a certain instrument and machine, 'Spectrochrome,' would cure any and all human ailments."

The highly respected Dr. Kate W. Baldwin literally blew the prosecution out of the water with her testimony. Near the end of her examination by the defense she said:

> We will always need the undertaker. We do not claim that we will not, but anything that is in human possibility to be put in a normal shape can be done, with Spectro-Chrome better than it can with anything else, and, in many cases, it is the only thing that would put the patient in a condition to function.[2]

The defense asked Dr. Baldwin if she was still using chromopathy, to which she replied: "I use practically nothing else."

In summary, Dr. Baldwin said:

---

2   When Dr. Baldwin mentions Spectro-Chrome, she is referring to chromopathy (chromo-therapy).

If the doctors stopped learning when they came out of medical college, they would know mighty little; they would be of mighty little good to the community, less than they are as it is.

In cross examination by the court, Baldwin was asked if she had abandoned the conventional practice of medicine. She replied:

No, I have not abandoned it. I use it if I have to, but I will not use it as long as I can get Spectro-Chrome. If I was cut out somewhere where I could not get Spectro-Chrome, I would have to go back to the next best thing.

"You apparently have been convinced," the prosecution asked, "through the teaching of the colonel and other things, that Spectro-Chrome surpasses medicine?"

"I have."

Prosecution: "So that in your mind you have practically abandoned the practice of medicine?"

Dr. Baldwin: "Only in the matter of emergency would I use the old methods of treatment."

Dr. Baldwin shocked the court when they asked her about the treatment of cancer:

In many cases of cancer, ... if there has not been too much destruction of tissue, Spectro-Chrome will cure it, it will build up the tissue. If it has to come to operation and there is a great deal of destruction of tissue, it will simply make them comfortable, for the rest of their life. It will make them comfortable so that they can enjoy the rest of their life to a certain extent, without doping them with opiates.

The prosecution continued to hammer away at the point that Dr. Baldwin had "turned her back" on conventional medicine.

She replied without hesitation: "I have. I would close my office tonight never to see another sick person, unless it was an emergency, if I had to go back to old style medicines and give up Spectro-Chrome."

It is important to keep in mind that this testimony is not from your country doctor, but by one of the outstanding physicians of her time. Dr. Baldwin, like **Denshah Ghadiali**, was persecuted and asked to resign from the staff of her hospital for supporting chromopathy.

The following chart shows the immense range of therapeutic efficacy of chromotherapy. It is a tragedy that deliberate suppression by vested interests have prevented this therapy from being used to help the suffering of mankind.

Chromopathy Therapeutic Wavelength (Color) Guide*

Scarlet	Magenta	Purple	Violet	Indigo	Blue
Arterial stimulant	Suprarenal stimulant	Venous stimulant	Splenic stimulant	Parathyroid stimulant	Antipruritic
Renal energizer	Cardiac energizer	Renal depressant	Cardiac depressant	Thyroid depressant	Diaphoretic
Genital excitant	Diuretic	Antimalarial	Lymphatic depressant	Respiratory depressant	Febrifuge
Aphrodisiac	Emotional Equilibrator	Vasodilator	Motor depressant	Astringent	Counter-irritant
	Anti-arrhythmic	Anaphrodisiac	Leucocyte builder	Sedative	Anodyne
Vasoconstrictor		Narcotic		Pain reliever	Demulcent
		Hypnotic		Hemostatic	Vitality builder
		Antipyretic			
		Analgesic		Phagocyte builder	

*Spectro-Chrome Encyclopedia, Malaga, N.J., 1939.

Turquoise	Green	Lemon	Yellow	Orange	Red
Cerebral depressant	Pituitary stimulant	Cerebral stimulant	Motor stimulant	Respiratory stimulant	Sensory stimulant
Tonic	Disinfectant	Thymus activator	Alimentary tract energizer	Parathyroid depressant	Liver energizer
Skin builder	Antiseptic	Antacid	Lymphatic activator	Thyroid energizer	Irritant
	Germicide	Antiscorbutic	Splenic depressant	Antispasmodic	Vesicant
	Bactericide	Laxative	Digestant	Galactagogue	Pustulant
	Detergent	Expectorant	Cathartic	Antirachitic	Rubefacient
	Muscle builder	Bone builder	Cholegogue	Emetic	Hemoglobin builder
	Tissue builder	Anti-arrhythmic	Anthelmintic	Carminative	
			Nerve builder	Lung builder	

Chapter 20

Military Use of Ultraviolet Irradiation

Through three major wars – World War II, Korea, and Vietnam – the American military has ignored the utility of treating troops with ultraviolet therapy. Since the beginning of recorded history, armies have been incapacitated or handicapped because of illness among the troops. When you amass large numbers of human beings closely together, as is inevitable in a military situation, there is going to be a lot of lost man-hours due to disease. History reveals to us many examples of how wars have been lost because of decimation through infectious diseases.

Without some understanding of the cataclysmic effects that infectious diseases have had, we are not able to get an overall picture of where we are today on this disastrous continuum of man's constant battle with disease and its intermingling with man's battle against himself. It would have been utterly impossible, for instance, for Cortez and his band of marauders to have conquered the highly religious and fanatically dedicated Aztecs of Mexico without his great secret weapon – smallpox. In Peru, history was repeated when Pizarro's little band of roughnecks conquered the Incas with crude weapons and a numerically inferior force, but with the same secret weapon. In fact, the reigning emperor died of smallpox while on campaign, and his designated heir also died, leaving no legitimate successor. Because of the breakdown in leadership due to disease, civil war ensued and Pizarro, with his brigands, was able to enter Cuzco, the capital, plunder its treasure, and take over the nation with practically no military resistance.

Even more devastating than the epidemic's effect on the bodies of the Incas and the Aztecs was the effect that it had on their minds and morale. It seemed perfectly obvious that these men who did not contract the disease were being favored by the gods, and, therefore, resistance was useless. It also became evident that their religion was superior to the native religion. This new God was far superior in that the disease only affected the natives and not the conquerors. This belief was, of course, reinforced by the Christian priests, who also sincerely felt that God was on their side. Consequently, the Inca and Aztec religions quickly disappeared.

In the time of David, it was said that 70,000 perished in one day. It is hard to imagine, even with our understanding of microbiology, what pestilence could have caused such a devastation. However, with modern technology, it would certainly be possible to kill 10 times 70,000 in one day with an attack of botulism toxin or a genetically altered, antibiotic-resistant, pneumonic plague organism.

One of the first recorded epidemics which affected the outcome of a war was the Athenian epidemic during the Peloponnesian wars which was described in the second book of the history of Thucydides. In the summer of 430 B.C., large armies were camped in Attica, the hinterland around Athens. Consequently, the people of the countryside swarmed into Athens, creating very crowded conditions. The Plague of Athens, probably typhus, had a profound effect on history. The citizens were seized rapidly with headache, red eyes, inflammation of the tongue, sneezing and cough; then came acute intestinal symptoms with vomiting, diarrhea, and excessive thirst. This was usually followed by delirium with the patient usually dying in about nine days.

Because of this pestilence, Pericles did not attempt to expel the enemy, who was ravaging Attica. Athenian life was completely demoralized with anarchy and extreme lawlessness. As Thucydides said:

> They saw how sudden was the change of fortune in the case both of those who were prosperous and suddenly died, and of those who before had nothing but, any moment, were in possession of the property of others.

Although Athens was totally devastated and helpless, the Peloponnesians left Attica quickly, not out of fear of the Athenians, who were locked up in their cities, but because they feared death from whatever disease God had wrought upon the Athenians. The

Greeks, in turn, lost their battle with the Peloponnesian fleet along the Peloponnesian coast because they were simply too sick to carry out their objectives. It is believed that this epidemic, also called the "Plague of Thucydides," was typhus fever. But it may have been smallpox.

Britain's history was very clearly influenced by a medieval pandemic, the nature of which is not known. In the year 444 A.D., a terrible pestilence struck Britain, an epidemic which was largely responsible for the historically momentous conquest of Great Britain by the Saxons. In fact, the Saxon chieftains were actually called upon for assistance because the Britons could no longer resist incursions from the north due to the depletion by plague of their fighting forces. So the present character of the British people, their mores, their customs, and their architecture, was greatly determined by a plague which occurred 1500 years ago.

The outcome of war turning on the fate of a tiny organism too small to be seen by the naked eye will occur again. Organisms rapidly develop a resistance to antibiotics. Rats, a common carrier of many illnesses, develop resistance to various types of rodenticides. In the next war, we can anticipate that the final outcome could be determined by which side has the best means of combatting this "second enemy."

Added to the calamity of natural infection is the problem of secondary infection from wounds. The specter of gas gangrene, for instance, hangs over any military operation. This dreaded disease comes from an infection with bacteria from the group, clostridium, which causes gas-forming organisms leading to dreadful, foul-smelling, rapidly fatal infection, if not properly treated.

In times of chaos and deprivation, as we might well be facing in a future war fought on our own soil, ultraviolet irradiation of blood, which is extremely inexpensive and simple, could be a deciding factor in whether the war is won or lost. This treatment would be especially applicable in virus epidemics among troops. As we have documented, almost all viral diseases resistant to all other known forms of therapy rapidly succumb to ultraviolet irradiation. Any type of bacterial infection, with the exception of bacterial endocarditis, will also quickly respond to this treatment, obviating the need for cumbersome and expensive antibiotics.

In view of the fact that we may face biological warfare in the future (see *AIDS – The End of Civilization*, chapter 6), it is of the utmost importance that the American military closely examine this great boon to mankind: extracorporeal ultraviolet irradiation of the blood.

Chapter 21

Russia Sees the Light

I have always felt, and experience has usually confirmed, that Western science is generally a decade or more ahead of Russia and its former satellites. But necessity brings out the best in the scientists of any country, including those in Russia.

If you can't afford antibiotics for the treatment of infectious diseases, then what do you do? There are two choices: Let people fend for themselves when they get a serious infection, which would end disastrously in an epidemic, or devise a cheap and effective alternative to antibiotics.

Russian physicians have always been a rather remarkable lot. No matter what kind of oppressive circumstances they must endure, they continue their clinical work and research with great enthusiasm -- on a monthly income that wouldn't support one of our welfare recipients for more than a day. Their performance during the 900 days of the siege of Leningrad (World War II) is a perfect example of their ability to perform under the worst possible conditions.

The heroic, almost superhuman defense of Leningrad against the Nazis would not have succeeded if a major epidemic had broken out within the city. History is replete with examples of armies, and thus nations, going down to defeat due to the ravages of disease.

From Pulkova Hill you can see St. Petersburg, 20 kilometers away. It was inspiring for me to stand there, realizing that Hitler's generals, looking through their binoculars, could see the street cars pushing through the deep snow of the city. Although the town had been reduced to mostly rubble, they were never able to conquer it.

During this terrible two-and-a-half years, Russia experienced some of the worst winter weather in its history, and over 600,000 of its non-combatants fell to starvation. But in spite of all this, not one major epidemic was allowed to add to the misery of the courageous people in the city of Peter the Great!

This singular medical and public health achievement was due primarily to the work of the staff of the Pasteur Institute of St. Petersburg, (a department of the Soviet government – not affiliated with the Pasteur Institute in France). The staff of only 55 scientists and personnel resisted the temptation to eat their laboratory animals, although they were in a state of near-starvation, and continued their research and medical defense of the city against the ravages of pestilence which would, almost certainly, have been fatal to the resistance. Every day these courageous defenders would go out into the city, which was under almost constant bombardment, to inspect and contain areas of infectious disease knowing that they may not return.

No one has ever doubted that Russian physicists and engineers have the ability to compete with any in the world. Their problem is that much of the industrial hardware made in the old Soviet Union did not work very well. For instance, there are antiquated nuclear power plants such as Chernobyl (and some worse), ecological disasters caused by politically-motivated programs, such as the total destruction of the Aral Sea. Besides this, there have been an appalling number of train, plane and submarine disasters due to faulty design, inferior workmanship, poor maintenance and poorly-trained support personnel.

All of these disasters and more were caused by bureaucrats making decisions that should have been left to engineers. For 70 years, a scientist dared not contradict his "betters" – members of the Communist *nomenclatura* who were ready and willing to destroy you for the slightest deviation or sign of disagreement with the established policy. These cowards and madmen gave a bad reputation to thousands of brilliant and dedicated scientists.

The Russian work in the field of photobiology is an excellent example of the old adage that necessity is the mother of invention. There was such a desperate need for an alternative to expensive antibiotics that even the bureaucrats left the scientists alone to do the necessary technical innovation and the clinical trials on human patients. Years of unnecessary animal experimentation were bypassed because the Russian doctors knew, from the work of the American engineer, Emmett Knott and his medical colleagues in the early 20th century, that UV irradiation of blood was a very safe procedure.

Before going to Russia in 1991 to investigate the status of phototherapy in that country, I was skeptical and thought I would find a technology in its infancy at best. I had heard rumors that they were doing work in the field of light therapy but I did not expect to find much, other than possibly a few ideas as to how we could improve our own instrumentation. I was prepared to enlighten my Russian colleagues on the subject of photobiology. My basic mission, as I viewed it, was to have a good time and impart some wisdom to my sincere but backward Russian colleagues. I was wrong and the flow of knowledge went in the other direction – from the Russian doctors and engineers to the smug American.

Russian medicine has been employing UV blood therapy long enough (20 years) to have overcome most of the resistance from orthodox medicine in their country. Any lingering doubts have been dispelled by the consistently good results they have obtained. In an area of the world where it is not considered witchcraft to hold a two-day session on *The Physics of Homeopathy*, resistance could not have been great in the first place. Russian and Eastern European scientists don't suffer from the self-satisfied dogmatism of the West that cherishes the delusion that our available knowledge is somehow infallible and final.

Although the intent of the light therapy research done in the former Soviet Union was directed to finding a substitute for antibiotics, it quickly became apparent to Russian doctors that the therapeutic applications of light to blood went far beyond infectious diseases and, in fact, UV light was only one of a vast array of energy applications that could be applied to blood for infections and other pathological conditions.

In the field of surgery alone, doctors of the former USSR have immense experience in the use of UV phototherapy. Over a half million procedures have been performed in over 100,000 patients. Surgeons Kutushev and Chalenko, of the City Center for the Fotomodification of Blood, St. Petersburg, reported that the use of UV light therapy cut by 50 percent the number of complications, and the necessity to use antibiotics, in severe trauma cases.

In the past ten years, these two surgeons have successfully treated over 3,000 patients with severe trauma, using UV blood irradiation. These cases ranged from crushed kidneys to extensive bleeding into the chest or abdominal cavities. They report the successful use of the patient's own blood when collected from chest or abdominal wounds. The blood is irradiated and then given back to the patient as an "autotransfusion." This retransfusion of the patient's own blood has the advantage of eliminating the risk of incompatibility reactions to a donor's blood, a common and often serious complication of blood transfusion. [1]

Another form of trauma, often, agonizing for the patient and therapeutically hopeless, is the tragedy of extensive, third degree burns. We discussed the use of external blue light for these patients in Chapter 14 in the remarkable case of little Grace Shirlow. (If you skipped over this incredible case report, go back and read it – it's even worth a second read.)

Dr. V.M. Novopoltzeva and his co-workers, from Mordovsky University, Saransk, Russia, reported in 1992 on 16 cases of severe third degree burns, some of them covering as much as 69 percent of the body surface. They employed an instrument similar to that now being employed by clinics worldwide, called Photolume III.

The surgeons made the following observations:

1. The patients' "common state" improved almost immediately after re-infusion of the UV-irradiated blood. Their appetite also improved markedly.
2. The severe pain subsided and they were able to stop injections of narcotics in many of the patients.
3. Because of these favorable clinical changes, the patients would often fall into a deep sleep for the first time since their accident.
4. The protein content of the blood plasma usually increased after the first infusion of irradiated blood – a very good sign in burn patients as protein loss is one of the major problems in these cases. [2]

One of the most interesting Russian studies was done on a mysterious, but common, condition called atopic dermatitis. The research was reported by Dr. Gromov and associates from the Vladivostock Institute of Medical Climatology.

As with many diseases, we know a lot about this pesky skin condition but we don't know enough to cure it or to even consistently

---

1 Saransk, 1992, pp 9-12., in Russian.
2 Saransk, 1992, in Russian.

treat it successfully. The condition can present with different appearances on the skin, as with many skin diseases, so we will not elaborate further on the clinical picture. Suffice it to say it is a skin malady frequently seen and often unsatisfactorily treated.

Their cases were especially interesting because they illustrate one of the few disease categories where the UV blood therapy can actually make the condition worse, if the treatment isn't administered cautiously. The investigators found that "standard treatment," i.e., two or three procedures a week, using 100 to 150 mls of blood, was "badly tolerated" by patients with severe atopic dermatitis, and would cause "significant relapse" of the disease with increased itching and the proliferation of blisters.

However, the investigators noted, even the most severe cases of atopic dermatitis should be treated with UV blood therapy but with an attenuated program – more time between treatments, shorter treatments, less volume of treated blood and less quantity of UV light.

With these caveats, the authors reported a 17 percent decrease in time spent in the hospital by their patients and the patients had a better "distant" result, i.e., they remained free of symptoms longer than controls not given light therapy over an observation period of six years. [3]

Some researchers have reservations about treating tuberculosis with UV light therapy, feeling that there may be a danger of dissemination of the TB germ due to a breaking down of walled off areas in the lungs. The treatment may be "too quickly effective" and thus cause a massive and overwhelming spread of the TB germ throughout the body. There may be some justification for this fear and, for that reason, TB patients should be treated in the same cautious manner as patients with atopic dermatitis. But the American experience in the mid-twentieth century confirms that photoluminescence is the treatment of choice for tuberculosis.

A Ukrainian study reported on the use of UV light blood therapy in cases of ear, nose and throat disease (ENT). Dr. Filatov and his co-workers at the Kharkov Medical Institute reported on 173 patients with a variety of ENT problems including chronic

---

3 Saransk, 1992.

sinusitis, chronic middle ear infection, nasal furuncles (sores inside the nose) and 12 cases of "rhinogenic sepsis," a general blood poisoning of the body from a nasal infection, often leading to death.

Their best and most dramatic results occurred, as our early American investigators reported, in cases of blood poisoning. The Ukrainian doctors noted "rapid disappearance of intoxication and fever." (Does that sound like a voice out of the past?) As in early American cases, they reported "extremely good effect" in patients who could not be treated with antibiotics because of hypersensitivity to the drugs and, they reported, many surgical procedures were avoided because of photohemotherapy.

It was noted in cases of sinusitis that the duration of remission was extremely long, usually more than three years. This will be good news to the tens of thousands of people who suffer from this chronic and often debilitating condition.

Another interesting and significant result of their study was the decrease in healing time observed following surgery of the ear drum when UV blood irradiation was used after the surgery. "Epithelization," healing, took place in *half the time* as compared to cases in which UV light was not used.

Russian scientists have developed laser beams for treatment of blood and are now working on X-ray and EMR applications. The X-ray technology has been used successfully in the treatment of leukemia in the United States. I was told an amazing story (by a radiologist who wants to remain anonymous) of recovery by six patients with leukemia, who were treated by X-ray blood irradiation.

He said the results were unequivocal and remarkable. He reported the cases to some of his colleagues who, he said, "started treating me as though I were a criminal." He quickly turned his back on any further investigation in that area and admitted to me, sheepishly, that he has felt guilty ever since. That will not happen in Russia. Although you find some of the same jealousies and back-stabbing among Russian scientists that you find in our country, there is clearly more of a spirit of open-mindedness there.

In Chapter 14, I stated that wavelengths of the color (visible) spectrum of light would someday be used in the treatment of disease as a refinement of Densha Ghadiali's external use of various colored lights on the body. The section of the book on color therapy was written three years before I started my investigation of Russian phototherapy. Color therapy has been developed by

Russian scientists to the point that we can now use it for the treatment of disease and test Ghadiali's claims directly on blood.

The color green, as an example of the broad opportunities presented here, is the color of balance and tranquility, according to the writings of Ghadiali. Green is located in the middle of the color spectrum with red and violet at extreme opposite ends. Would it not be better to treat anxiety, depression and other neuroses with green light radiation of blood rather than dangerous and mind-altering drugs?

With convincing evidence accumulating in the field of infectious diseases, other specialists began to take an interest in photobiology. Almost every medical specialty in Russia is now involved in the continuing research of light therapy. The body of scientific clinical investigation in the light field in Russia is now enormous, and includes the disciplines of surgery, neurology, rheumatology, pulmonology, gynecology, dermatology, toxicology, cardiovascular diseases, as well as all branches of infectious diseases – bacterial, viral and protozoal. Over a hundred hospitals in the former Soviet Union have had hands-on experience with the Russian phototherapy techniques.

Bacterial endocarditis, an infection of the valves of the heart, is one of the most dreaded infections encountered in medical practice. Even with the most advanced medical therapy, patients still may die from this condition.

Endocarditis is apparently more common in the Ukraine, formerly a part of the old Soviet empire. Dr. Krishtof and his associates were able to treat an amazing 250 cases of this infection.

All of the patients they treated had undergone prolonged therapy with antibiotics and cortisone, with little effect. They remained highly toxic with high temperatures and laboratory evidence of immuno-deficiency. Because of the seriousness of their condition, patients received two to three UV treatments daily.

This intensive therapy with UV light was so successful that 43 of the patients were able to avoid surgical repair of their heart valves – a remarkable accomplishment. Those who had surgical repair, fared better post-operatively than would be expected from such serious surgery, and their hospital stay was significantly shortened.[4]

---

4 Saransk, in Russian, 1992.

Doctors from the Sklifosovsky Institute for Urgent Medicine studied the application of UV light therapy in chemical toxic states. Their series of 128 patients had organophosphate or psychotropic drug intoxication.

The recovery from coma was, on average, twice as fast as was seen in patients not receiving phototherapy. The patients receiving UV blood irradiation had 50 percent fewer complications, such as pneumonia. Also, there was a 40 percent reduction in mortality in the UV-treated group. [5]

A report from the Krasnoyarsk Cancer Center on the use of UV blood therapy in conjunction with surgery in cancer of the colon and rectum is remarkable. It should be repeated by other cancer centers without delay. Their results were so astounding that they will not be accepted, and probably shouldn't be, until a rigorous clinical trial is conducted to confirm or deny their work.

Twenty-two patients with cancer of the colon and rectum were treated after surgery for their cancer starting on the first post-operative day. They were treated four times a day, which is much more often than generally recommended by specialists in this field. They found:

1. Narcotics were often not necessary following surgery or they could be discontinued on the second day.
2. There were no cases of adynamic ileus, the paralysis of the intestines so often seen following abdominal surgery.
3. Post-operative infections were seen in only ten percent of the patients compared with 30 percent in those patient not receiving light therapy.[6]

The treatment of cancer is an intriguing and exciting area of research in phototherapy. We have reported to you the work of Olney on the treatment of cancer, and the impressive results that Edelson has had in the treatment of cutaneous T cell lymphoma. Russian investigators have also had success in cancer therapy using UV light in both experimental animals and humans.

Dr. Igor Dutkevich from Hospital 15, St Petersburg, did research on mice with melanoma and lung cancer. Dr. Dutkevich

---

5 Saransk, 1992.
6 Saransk, 1992.

made a remarkable discovery when he compared the use of blood from another animal with the effects seen when the animal's own blood was used. When the blood from another animal was irradiated with UV light, then injected into the cancerous mouse, the cancer became immediately worse with the rapid dissemination of the cancer throughout the body. When the blood of the cancer-afflicted mouse was irradiated and re-injected, the cancer was inhibited.

There is no known mechanism for this interesting, and undoubtedly important, phenomenon. Perhaps some auto-immune mechanism is involved when the mouse's own cells are used.

Dutkevich also studied the use of phototherapy as an adjunct in cancer surgery. Seventy-eight patients with lung cancer were treated with UV irradiation of their blood prior to surgery. He found that the patients receiving phototherapy had shorter stays in the intensive care unit, as compared with 121 patients who did not get the therapy. They also had a decrease in the frequency of post-operative complications (from 20 percent to 11 percent) and a marked decrease in mortality from 5.8 percent to three percent. Dr. Dutkevich found similar improvements in patients with stomach, kidney and bladder cancer.

Still, Dutkevich and his associates give a clear warning concerning the treatment of leukemia, pointing out that too high a dosage of UV can produce serious complications including an aggravation of the disease. The use of much lower doses, as in the treatment of atopic dermatitis, are recommended.

Hypertension, or high blood pressure, is an area that was not investigated by the early American pioneers in blood radiation. A number of hospitals in Russia are investigating this important disease and their preliminary results are encouraging.

At the Azerbaijan Republican Hospital, in Baku, 34 patients were treated with UV extra-corporeal light therapy in combination with anti-hypertensive drugs. These patients were chosen for one or more of the following reasons:

1.   Their pressure was extremely high.
2.   The drugs were not effective in lowering the pressure.
3.   The patient was intolerant to anti-hypertensive drugs.

The patients were given a very intensive treatment regimen having received five to ten treatments every other day.

Some lowering of the pressure was often seen with the first treatment. At the end of the series of treatments, some patients experienced a 30 percent decrease in systolic and diastolic pressures. Symptoms of high blood pressure, such as headache, dizziness and chest pain usually disappeared and the remissions were prolonged for at least ten months with no evidence of a progression of their vascular disease. Use of drugs was drastically reduced which added to the symptomatic relief of the patients. [7]

A group of St. Petersburg physicians studied the effects of photoluminescence on 145 patients with severe blockage of the arteries to the heart, who had suffered a previous heart attack. This study was particularly impressive because they chose only patients who had not responded well to conventional drug therapy.

"Significant improvement" was registered in 137 of the 145 patients treated as compared to control patients who did not receive light therapy. Pain was quickly relieved and analgesics were often discontinued. The dosage of heart medications, such as beta-blocking agents, were reduced in most patients and the attacks of angina were less frequent than the controls.

The authors attributed the good results to a dilation of blood vessels to the heart, normalization of blood-clotting mechanisms, increase in the oxygen-carrying capacity of the blood, and improvement of oxygen uptake in tissues. They hypothesized that the effect of the light on the blood and the blood vessels was due to photodestruction of some, yet to be identified, harmful blood proteins. [8]

Dr. Lev Kukui of the Lenin Hospital has had extensive experience in the treatment of coronary heart disease with phototherapy. In a 14-year span, Kukui and his co-investigators treated 256 patients with serious heart disease. Ninety-five percent of their patients showed at least partial amelioration of symptoms and only nine percent of them could not return to their usual professional activity. The dosage of nitroglycerine needed to prevent chest pain was reduced to three tablets a day as compared to ten tablets a day by those patients not receiving phototherapy.

Using red laser for blood irradiation, they were able to prevent heart attacks in 90 percent of their patients with severe

---

7 Saransk, 1992.
8 Leningrad: Nauka, 1986, in Russian.

angina pectoris, whereas only 30 percent of those patients not receiving the therapy avoided a heart attack. Arrhythmia, or heart beat irregularity, was treated successfully in 81 percent of the light-treated group but only in 33 percent of the untreated group.

A. Levin and his associates studied the effects of UV blood irradiation on blockage of the arteries of the legs, a common problem in diabetics and heavy smokers. They reported positive results in eight of eleven cases with significant relief of pain, less dependance on pain-killers, better sleep, better appetite, and quick healing of ulcerations caused by poor circulation.

One of the major problems in surgery remains the inflammation and blockage of the veins of the legs following a surgical procedure. This "thrombosis," or blockage of the leg veins, prolongs the hospital stay and can lead to a fatal blood clot traveling to the lungs. Dutkevich and his associates reported that 10.3 percent of surgical cases in their series developed some degree of thrombophlebitis or thrombosis following surgery, if not treated with UV light therapy. But *not a single case* developed venous complications if they had been treated with UV light therapy prior to or after surgery.

A more drastic procedure is the use of laser light in various color frequencies directly into a vein or artery. This type of therapy, as far as I have been able to determine, has not been reported in American medical literature. Although it sounds like "star wars" medicine, the procedure is relatively simple. We observed this interesting therapeutic technique at the clinic of Drs. Dutkevich and Marchenko at Hospital 15, mentioned above.

To get to Hospital 15, you have to drive halfway to Siberia, or so it seems. We took the subway to the end of the line going toward Finland and then a bus bulging with four times as many riders as seats. At the end of the bus line, we walked on slippery, frozen streets, in sub-zero weather, for about a mile to the concrete maw that was the entrance to Hospital 15. (I've had to suffer a lot to write this book, steaming jungle heat in Africa and Arctic, blue-toe weather in Russia. But *you* probably don't care.)

As we approached the entrance to the ten-story hospital, three reeling, terminally-drunk men were leaving the building, heading for the vehicle depot behind us. It was ten o'clock in the morning and they were going to work. They were *ambulance* drivers. (Policemen are sometimes seen in a similar condition.)

We carefully negotiated our way around these protectors of the nation's health, and entered Hospital 15.

As with many of Russia's scientific institutions, the elevators were "temporarily in a state of repair," which means permanently disabled, and so we walked up seven flights of dirty, unlighted and freezing-cold stairs to the doctors' offices.

I would like to make some parenthetical observations here about the paradox between Russian, cutting-edge scientific investigation, and the inability of the system to maintain even a modicum of modern infrastructure for their pauperized scientists. Then we will return to Hospital 15.

I visited the clinic of Dr. Aza Rockmonava, St. Petersburg's AIDS treatment specialist, at the Bodkina Hospital in the center of the city. After trudging up the usual multiple, dingy stair cases, we were confronted with a massive door made of a bamboo-like material that would have been right at home in many shacks I have visited in central Africa. In fact, the conditions at most of the *best* Russian hospitals are only a little better than what I encountered in Africa. The conditions in the hinterland of Russia are *worse* than in much of Africa. The reason for this is the Africans have had the advantage of well-equipped Christian missionary hospitals for the past 50 years, something the communists would never allow.

There is a laser research institute in the center of St. Petersburg that is an even more dramatic illustration of brilliant scientists working in buildings that are little better than cow sheds. The center is located in a 19th-century structure under which the planners of the Brave New World decided to build an extension of the St. Petersburg subway. They didn't reinforce the excavation adequately which led to a sinking and consequent buckling of the middle of the building, making the middle third uninhabitable. In typical Russian socialist fashion, the middle portion was boarded up and it was business as usual in the surviving ends of the building. The scientists, accustomed to catastrophe, continue their research on the effects of laser light on biological systems as enthusiastically as ever. As you have probably guessed, the elevators don't work there either.

One day a colleague and I were ascending in a hospital elevator that did work, all be it with great groaning, lurching and some strange clicking noises that sounded like electricity being set free to go where it might, along the line of least resistance.

(I didn't touch anything.) I asked my companion how often the elevators got inspected and he replied that he had never heard of one being inspected – EVER!

Now, back to the concrete monstrosity called Hospital 15. Drs. Dutkevich and Marchenko are doing creative work here in spite of their very limited resources. In their technique, an intravenous drip is applied in the usual way to a large vein in the arm, near the elbow area. Then a stiff, hair-like fiberoptic thread is introduced into the tubing and run down to the opening of the needle. The laser is then turned on and the light courses down the thread and radiates into the blood. You can easily see the light glowing under the skin at the point of the needle and a few inches beyond. Those scientists who say light can only penetrate the skin for a few millimeters only have to witness this procedure once to realize how wrong they have been. By directly injecting light energy into the blood, the entire blood supply can be irradiated in one treatment session!

There is a great deal of debate and antagonism concerning the intravenous method of Dutkevich/Marchenko. There have been many debates concerning the safety and effectiveness of the technique. But now that capitalism has entered the medical scene, as well as most other areas of Russian life, there is a fierce competition between the advocates of the various techniques. Some, new to the spirit of free enterprise, don't handle it very well. One surgeon in St. Petersburg calls all of his competitors "govno." We will have to leave it untranslated and you will not find the word in your Russian dictionary.

We observed the intravenous phototherapy technique on four patients being treated at Hospital 15. They were perfectly comfortable and there was no evidence of toxicity during the course of the therapy. Although the critics of the method claim, most emphatically, that it is dangerous and fraught with many side effects, Dutkavich/Marchenko claim they rarely see *any* side effects and absolutely *none* of a serious nature. The critics cited no examples of adverse reactions to the method but were convinced of the dangers none the less.

Dutkevich and Marchenko will be reporting on their work in a subsequent textbook on photoluminescence that they hope to have published next year.

You will note that most of these Russian reports come from a scientific conference that was held at Saransk, Russia. Because of the historic nature of the conference, and the voluminous amount of light research reported from it, I asked one of my colleagues if he thought I should visit the city. He replied: "Not if you can possibly avoid it." So I have avoided it.

Doctors at the Nizhny Novgorod Medical Institute studied the use of intra-aortal light therapy in the treatment of peritonitis, an often fatal infection of the lining of the intestines and the abdominal wall. This method of introducing the light into the aorta, which comes directly off the heart, is far more complex than the intravenous method described above, but has the advantage of energizing the blood immediately before it enters the portal circulation which supplies the intestines.

The doctors reported the following advantages to patients treated with intra-aortal light therapy as compared to those not receiving such therapy:

1. The period of post-operative toxicity was shorter.
2. The period of post-operative bowel paralysis, which always occurs following this type of surgery, was shorter.
3. The hospital stay was cut by 35 percent.
4. Post-operative mortality was improved.
5. Improvement in blood parameters, such as T and B cell counts, immuno-globulins and immune complexes, began to improve the day after surgery. [9]

Another series of cases using laser light directly in the blood stream was reported by Drs. Krasilnikov and Karpuhin in 1992. They treated 52 patients with acute intestinal obstruction with intravenous irradiation of blood after their obstructions were corrected by surgery.

The patients were treated daily and demonstrated a quick response. This was evidenced by the disappearance of pain, easier breathing and improved laboratory values such as a decrease in circulating immune complexes, a decrease in the sedimentation rate and a decrease in the creatinine.

These doctors used another innovative approach that requires further investigation. During surgery for purulent peritonitis (an infection of the lining of the intestines and the

9 Saransk, 1992, in Russian.

abdominal wall with pus formation) they irradiated the bowel and abdominal cavity through the surgical incision, placing the laser light about five inches away from the target area. An area about four inches in diameter was exposed to the light. This bathing of the intestine with laser light was continued after surgery by leaving a fiber optic strand in place through the surgical wound.

This unique procedure paid off with a dramatic decrease in the death rate from this often fatal condition. The mortality decreased from 16 percent to nine percent. This technique, if confirmed by other investigators, will some day become a standard adjunct to all surgery involving infection. [10]

UV blood irradiation has a wide application in veterinary medicine. I, along with a veterinary colleague from Texas, tried to get American veterinarians to utilize photoluminescence for infections in animals. Much to my surprise and disappointment, I found that the vets were just as timorous and resistant to new therapies as the MDs.

Although we assured them the treatment was safe, they were not willing to try it on animals in their practice as they took all direction from the university veterinary schools. University investigation requires mountains of government or pharmaceutical money and years to complete. American veterinarians, like American MDs, have lost confidence in their ability to do their own research, or they just don't want to take the time away from a routine that is accepted and lucrative.

If you don't do anything innovative and out of the mainstream of medicine, you will not be criticized, you will not be punished, and you will make a good living. Veterinarians used to be quite independent and they resisted the conformity to which MDs fell prey. But drug company influence, governmental regulation and the commercialization of their profession has now caused them to also succumb to the paralyzing effects of the modern medical industry.

Russian veterinarians have not been so timid. In the late 1980s and up to the present, Russian vets have done impressive work in the field of photoluminescence.

---

10 Saransk, in Russian, 1992.

Newborn calves often die of diarrheal disease. Research proved that many could be saved by only two treatments with UV-irradiated blood injected into the muscle.

Acute respiratory diseases were treated successfully in a similar manner and the therapy required only half as long as conventional drug therapy for recovery of the animals.

In threatened epidemics, it was found that UV-irradiated blood could be given from another animal *en mass* to the threatened herd, and effective control of the infection could be achieved. This can represent a large saving in the use of prophylactic antibiotics. [11]

We are indebted to our Russian colleagues for the massive amount of work they have done in the field of photohemotherapy, so much so that we are including them in our dedication of this book to the true pioneers of the medicine of the future.

---

11 Samoilova, Russian Institute of Cytology, Acad. Sci., St. Petersburg, Russia.

Chapter 22

Postscript

When one reviews the incredible cases I've related in this book, and sees the remarkable simplicity, safety and efficacy of photoluminescence, one wonders why such an effective therapy was discarded. The only answer I can fathom is that the advent of the antibiotic era, with its potential multi-billion dollar profits, completely overwhelmed the minds of doctors. Through drug-company advertising, they became brainwashed into believing the miraculous power of antibiotics would cure all ills. It became impossible to even consider any other form of therapy.

Indeed, antibiotics are miraculous and have been a boon to mankind. But they are only for the elite, as most people in the world cannot, and never will be able to, afford them.

If one looks through medical journals today, it is obvious that doctors' minds are under the control of the drug companies. There are literally hundreds of thousands of dollars worth of advertising in practically every issue of every medical journal you read. These tens of millions of dollars spent by the drug industry are not expended for altruistic purposes. It has been estimated that the drug companies spend more money brainwashing doctors on drugs than is spent yearly on educating all the doctors in the nation.

Ultraviolet therapy represents a tremendous threat to the legal international drug cartel, and, for that reason, it has been suppressed. There will be vicious infighting and cries of quackery when this book is published, and there will be a concerted effort to ignore it – or destroy it.

However, with the recrudescence of a huge army of infectious diseases, and with the rapid deterioration of the effectiveness of many of the standard antibiotics, this conspiracy of silence will be broken; and ultraviolet irradiation – photoluminescence – will have its second day in court. The fact that the treatment is effective against toxins and viruses, where antibiotics are essentially useless, makes it all the more certain that extracorporeal ultraviolet irradiation will take its place in the forefront of treatment in the late 20th century.

In spite of the billions of dollars spent on research, development, and marketing of antibiotics, myriads of strains of bacteria, some old and some new, are threatening the health and the very lives of people all over the world, including the United States. Lethal viruses continue to prey on the human race and are rapidly increasing in virulence as more deadly viruses follow AIDS into the medical scene.

The most startling recent example of the awesome lethality of these new super bugs is the case of Jim Henson, the famous creator of the Muppets. Muppeteer Jim Henson's sudden death in May of 1990 shocked millions of children and adults who shared a special fondness for his wonderful puppet creations. Henson's death was unsettling on a different level, however. The bacterial infection to which he succumbed – a particularly virulent strain of group A Streptococcus – causes a newly recognized syndrome called toxic shock-like syndrome (TSLS), that can fell otherwise healthy people within hours of the onset of symptoms. It was astounding and very unsettling to have healthy Jim Henson die within three days of falling ill.

This lethal bacterium, which responds to none of the available antibiotics, has spread around the world and surfaced in England, Scandinavia, Australia, East Germany, Canada, and New Zealand, as well as the United States. Scientists are bewildered as to why this organism, which ordinarily causes strep throat in children, is suddenly causing an acute fatal disease in adults.

The importance of the Henson case from the standpoint of this book is that what killed Jim Henson so rapidly was the virulent toxin produced by the Streptococcus A organism. This toxin will not respond to any antibiotic. So even if the antibiotics had killed the organisms present, Henson would still have died from the toxin produced by the organism.

As you know from reading chapter 4, photoluminescence will have an immediate and permanent therapeutic effect on toxins and will clear the body of this deadly poison, usually within hours. I have no doubt that Jim Henson would be alive today and would be continuing his wonderful puppet artistry if photoluminescence had been immediately available at the hospital where Henson was taken.

A recent report, which *Science* called "chilling," described a new strain of shigellosis, a virulent and often fatal form of dysentery. The infection spread like wildfire through a Hopi Indian reservation in Arizona. The Shigella organism responsible for this epidemic was resistant to all the known antibiotics and cried out for the use of photoluminescence. But, unfortunately, this miraculous therapy is not available even in the big-city medical centers, much less on an Indian reservation.

As we were going to press with this book, the Wistar Institute made a report public that sent tremors through the medical community and is forcing the resurfacing of a question asked three years ago in my book, *AIDS – The End of Civilization*. Will the biological disaster of AIDS, and other deadly viruses, spell the end of Western civilization? The Institute announced that the Chronic Fatigue Syndrome, which is believed to affect ten million Americans, is caused by HTLV-II, a viral organism that causes hairy cell leukemia.

Keep in mind that there is no cure for any virus and no cure for any form of leukemia. What will America's doctors do if 10,000,000 young, mostly white, middle class citizens suffering with Chronic Fatigue Syndrome come down with hairy cell leukemia? With our experience in treating AIDS in Africa and the U.S., we are confident that we can control this new scourge with photo-oxidation. It is our best hope for preventing a disaster that would be unparalleled in the history of mankind, far surpassing the 500-year devastation of the bubonic plague.

When the bodies start stacking up – especially when people realize that the young and middle-aged people are dying, as well as the old – there will be "panic in the streets" and in the corridors of the nation's hospitals. The public outcry will force doctors to turn to this inexpensive, painless, safe, and effective therapy.

Our Light Research Continues

At the present time, we are only using one energy frequency, in the ultra-violet range. Future instruments, now under development in our laboratories, will use many different segments of the energy spectrum from UV to green (500 nm) to red (650 nm) through the infrared (700 nm). As mentioned in Chapter 19, all areas of the spectrum, from ultra-violet to red and infra-red have a place in the treatment of disease.

One of our most promising projects is in the use of the entire spectrum from 250 nm to 700 nm, i.e., full spectrum light just like from the sun, with photo-sensitive compounds added to the blood sample before exposure to the light. Using this technique, we think the body will then choose the portion of the activated sample it needs for healing.

At present, we have five substances that will activate the blood at various places on the spectrum, mostly at the ends – blue on one end and red on the other. Green (a neutral, balancing frequency at 500 nm) should be very important for blood irradiation therapy because green is the "cleansing frequency." Green is an antiseptic, a germicide, and a detergent. It is the governing wave (the normalizer of body functions).[1] It is the division line of the color spectrum with red at one end and violet at the other. So it is the balancer and the best frequency to use when the cause of the problem is in doubt. We are diligently searching for the proper compound, which may be an herb, a vitamin, or even an amino acid, that will activate cells when they are exposed to green light.

The prospects for blood therapy with color frequencies are immense. The frequency orange (620 nm) has a profound effect on the lungs. Would orange irradiation of the blood, combined with blue (400 nm) for infection, cure pneumonia without the use of antibiotics? Turquoise relieves inflammation of nerve endings. Would irradiation of the blood with turquoise reverse the terrible pain of shingles, which is an inflammation of nerve endings?

These are among the many questions we are seeking to answer in our work with light therapy. There is much exciting work to be done, and I hope many will join me in seeking to effect the widespread implementation of this miraculous technology, and to explore its possibilities even further.

Realizing those possibilities will mean the difference between health and sickness – and often, life and death – to millions of people. But it will mean the difference on an even larger scale. In the face of the AIDS epidemic, the very survival of our way of life is at stake. The key to the future health of our culture lies with widespread implementation of photoluminescence.

---

1 *Spectro-Chrome Encyclopedia*, p. 936.

Chapter 23

Epilogue

Now that you have read the preceding chapters regarding the almost miraculous benefits of photoluminescence, particularly in treating infectious disease, you are probably impatiently asking, *What is being done to make this treatment available today?*

If you are one of the millions of people with a loved one suffering from hepatitis, viral pneumonia, meningitis, AIDS, or a dozen other killer diseases — or, God forbid, you suffer from one of these ailments yourself — that question is not of just academic interest. It could literally be a matter of life or death to you.

It is my fervent hope that publication of *Into the Light* will help speed the day when photoluminescence is readily available throughout the world. And I am pleased to report that much is being done right now to help bring that day to pass.

Since returning from Russia, I have been invited to address two of the most important organizations of alternative physicians in the United States about my experiences with photoluminescence. In both instances, the audience was eager to learn more.

In the past eighteen months, I have traveled to Europe, Africa, North America, and the Caribbean, to meet with physicians and alternative therapists who expressed interest in learning about this remarkable healer.

Yes, word is getting out! Some doctors *are* listening to us ... some brave medical groups *are* looking into this incredible therapy. I hope we will soon be able to announce that some brave and pioneering physicians in the United States will offer photoluminsence treatments in their own practices, as part of an officially sanctioned study by an investigative review board.

As there is news to report, I'll do so in the pages of my monthly newsletter, *Second Opinion*. If you are not already a subscriber, I urge you to become one. Send $49 for a one-year subscription (foreign subscribers, add $13 a year for air mail delivery) to

Second Opinion, P.O. Box 467939, Atlanta, Georgia 31146-7939.

In the meantime, one of the most important things you can do is to encourage others to learn more about this remarkable therapy — including any doctors you know who are open-minded enough to listen.

Loan them your copy of *Into the Light*. Better yet, urge them to buy one themselves; that way, they will be much more likely to read it.

In conclusion, let me thank you for bearing with me for these hundreds of pages, as I tried to explain a revolutionary therapy that is really decades old.

I have spent more than ten years of my life confirming the almost miraculous results of photoluminescence. I was so eager to see it in widespread use that I endured a lifestyle that would be unbearable for most Americans — an Arctic winter in St. Petersburg, Russia, living and working with those wonderful, warm, and courageous people who are just beginning to recover from the devastation from 70 years of communist tyranny.

May God bless them all. As for me, I wouldn't change a minute of it.

William Campbell Douglass, M.D.
St. Petersburg, Russia
March 1993

Photo-Oxidation (Photox) Protocol for Physicians

There's bound to be a lot of resistance to ultraviolet blood irradiation as a valid method of treatment. At first blush, it probably seems to rank right up there with witch doctors and snake oil. But the concept of irradiating blood from an energy source is not new; since ultraviolet light is such an effective viricidal and bactericidal agent, its use as a blood irradiator has intrigued investigators for many years.

At first, attempts were made to irradiate thin layers of blood in petri dishes exposed to mercury-vapor lamps. More extreme (and cumbersome) methods were also tried, such as stirring a quartz-rod applicator of a lamp in a beaker of blood, and even placing quartz rods directly into arteries. All these crude methods failed and probably served to discourage research in the field of human photo-biology. Not until the work of Emmet K. Knott were the limits and optimal dosage of therapeutically applicable light determined.

The direct killing of bacteria in the bloodstream in cases of septicemia was the initial objective of blood irradiation with ultraviolet light. Beginning in 1923, ultraviolet blood irradiation was tested in over 40 dogs for efficacy in septicemia. Spanning a period of three years, the method was proven safe and effective in the treatment of bacteremia of almost any type.

It was first assumed that the entire blood supply of the dog would have to be treated to eliminate the infection. But to the investigators' surprise, although total blood irradiation returned a sterile blood, all the dogs died – apparently from shock. Various experiments, which varied the volume exposed to radiation and the time of exposure, revealed that only 1/16th of the blood supply had to be irradiated for excellent results (1.5 ml per pound of body weight.) The length of exposure to the light could be varied within wide limits. We use three minutes with the syringe method and, because of the larger volume of blood exposed and the more effective exposure, ten seconds with the continuous-flow, large-volume method.

An amazing amount of photonic energy is produced by the ultraviolet instrument. The photochemical effect has been found to be 925,000 ergs/cm^2/sec. This is about one hundred times the energy of ambient sunlight.

The photodynamic effect can be increased by the presence in the blood of certain photosensitive amino acids, herbs, dyes, and porphyrin derivatives. We use our own formulation called Photovite. Alkaline blood and a high blood oxygen also increase the effectiveness of ultraviolet blood irradiation.

Many theories have been suggested as to why photon energy transfer to the blood is effective – cross-linking of DNA, increased oxygen saturation of cells, immune stimulation and others. Any or all of them may be correct. But although the mechanisms of action of Photox are only partially understood, the astounding results speak for themselves.

Miley demonstrated an amazing increase in blood oxygen absorption in total blood following irradiation even when absorption was initially low. He found rapid increases of 60 percent, and an average increased oxygen absorption of 50 percent thirty days after irradiation. This remarkable continuation of high blood oxygen has not been explained but it must play an important role in the healing process that is observed.

In 1937 and 1938, Dr. Miley demonstrated that there was a decreased oxygen capacity in venous blood in many disease states. The rise in oxygen content of venous blood following ultraviolet irradiation, with no compensatory rise in hemoglobin or red-cell count which could possibly account for it, usually led to a dramatic improvement in the patient's condition and almost immediate pinkening of the skin.

Tachycardias due to anemia demonstrate a noticeable slowing of the heart within 15 minutes of blood irradiation with progressive improvement for several weeks. These changes were observed without drug intervention.

The relief of toxemia, such as in tetanus and botulism, remains a mystery. However, the lipid fraction of blood seems to be involved. Chylomicrons in normal blood are minute, separate particles with normal Brownian movement. In toxemias, the chylomicrons are clumped and there is no Brownian movement. After ultraviolet blood irradiation, the clumps, in vitro and in vivo, quickly break up and there is a rapid return to normal Brownian movement. The toxemia abates parallel with the improvement in the chylomicron's activity. One of the functions of chylomicrons may be to absorb toxins which would explain why the dispersal of the chylomicrons leads to detoxification.

Whatever the mechanism of action, the criteria for an acceptable therapy are met: (1) It is safe. (2) It is effective in a broad range of diseases. (3) There are no side effects. (4) It is relatively inexpensive.

Ultraviolet irradiation of blood has been approved by the FDA for the treatment of cutaneous T cell lymphoma. Thus the method is legal within the context of the FDA's definition of legality.

It is also legal from the standpoint of long (over 50 years) and continuous use by physicians in the United States as a commercially viable product before the present FDA was even in existence.

Length of Therapy

Acute illnesses, such as the common cold, minor bacterial infections not requiring drainage, pneumonias, toxic conditions such as venomous snake bites, and the childhood viral diseases, will usually respond within a few days of therapy on an outpatient basis. If using the intermittent ten cc syringe method, treat these minor illnesses twice a day. If the continuous flow, 1.5 ml of blood per pound body weight method is employed, a treatment every five days is adequate, more often for severe cases. With minor illnesses, one or two treatments by either method should be sufficient.

For chronic states, the treatment should be given three times a week until there is improvement and then according to the judgment of the doctor. It can be given daily, or even twice daily, with complete safety except in cases of atopic dermatitis and porphyria where a quarter dose (by varying the volume of blood, by the time of exposure of the blood, by the quantity of light allowed to energize the blood, which is controlled with the metal blind at the top of the photolume III, and by the frequency of the treatments).

The treatment protocol is essentially the same for all chronic disease states with the exception of the two above-mentioned conditions. With the above general rules, the doctor will vary the treatment as to dosage and frequency according to his clinical judgment.

A 23- or 21-gauge butterfly needle is placed in an arm vein or the external jugular vein. Ten cc's of whole blood is withdrawn into a pure quartz syringe (Careful – they are very expensive and quite fragile.) and put into the Photolume II instrument through a circular opening in the front of the machine. The syringe rotates slowly and is thus exposed to continuous ultraviolet light and other wavelengths of the photo-magnetic spectrum. The exposure time is three minutes.

The syringe containing radiated blood is then withdrawn from the Photolume and re-injected intravenously. The original IV is kept open during the extracorporeal ultraviolet irradiation by a heparin lock or by a slow hydrogen peroxide (0.015% solution) drip. As explained in the text, the increased oxygen level in the blood and tissues as a result of the H_2O_2 greatly enhances the effectiveness of the therapy.

Because of the mechanics involved, the syringe method requires about one and a half hours to complete. The IV hydrogen peroxide should be given slowly. This time can be used to repeat the ten-cc blood irradiations as often as desired. The continuous-flow, large-volume method can be done in 30 minutes.

The continuous flow system (Photolume-III) is a modification of the old Knott Hemo-irradiator invented in the late 1930s and applied to human disease in the 1940s. It consists of a multiple light source which approximates sunlight, a pure quartz window over a baffle which agitates the blood, a pump, a blood filter, and an attached IV stand. The pump controls the rate of flow.

The absorbed energy from the irradiation can have multiple effects. The absorbed energy may result in dissociation, fluorescence, simple dissipation of absorbed energy, or a translating of vibrational (light) energy to the surrounding cells when re-injected. Important chemical effects produced by ultraviolet radiations are decomposition, rearrangements, such as cis-trans reactions, and isomerizations, and sensitization (similar to photosynthesis in plants through the action of chlorophyll as the sensitizer). Ultraviolet rays also act as a powerful catalytic agent.

Photo-biology and photo-therapy have had little attention in the past 50 years but with the advent of the New Age of Pestilence, with new disease organisms such as AIDS, Palestine syndrome, HTLV-I, and chlamydia, and the counterattack of microbes we thought we had conquered, light therapy will become the treatment of choice for infectious diseases.

Biophotons and Biophysical Effects of Ultraweak EM Radiation

by Scott Hill M.E.E.
University of Copenhagen
(Postboks 1168, DK-1010 Copenhagen K)

Proceedings of International Symposium on Wave Therapeutics May 19-20, 1979.

During the last two decades considerable advances have been made in understanding the mechanisms involved when biological tissues are irradiated with electromagnetic waves of different frequencies. For high-powered radiation, thermal effects are the dominant reactions, but scattered reports over the years indicate that secondary, non-thermal effects may also occur. In the last ten years, the literature on non-thermal effects of EM waves has mushroomed enormously, indicating a renewed interest in this overlap area between medicine, electrical engineering, biophysics, and biology. Work on microwave and radar frequencies has been described by the Soviet researcher A.S. Presman, who reports that minimum dosage standards in the USSR have been set 10 to 100 times lower in the USSR than is common in the US for radar and radio operators.

A recent book by Soviet biophysicist Yuri Kholodov described work on neural effects of EM waves from DC up to gigahertz frequencies. The Canadian researcher M.A. Persinger discusses possible mechanisms for how an extremely weak extra-low-frequency

(VLF) EM field might affect bioregulatory mechanisms involving the whole-body, organ, and cellular levels.

EM waves (especially microwaves, but applicable to any frequency range), can interact with biological tissues through a variety of mechanisms. Certainly we know that light has a major effect on biological systems: the growing fields of photobiology and photochemistry indicate one way in which the investigation of the action of visible (and other wave bands of) light on biosystems has been undertaken. In the Soviet Union, there is a large interest in the effects of very weak light on biosystems. A corresponding effort is made to study the very weak light emitted by biosystems: bioluminescence and dark chemiluminescence in the visible, UV and IR bands, which has opened up an entire new field of low-energy biophysics. Today, special interest is being paid to the role of monochromatic, polarized light, including laser light, on biosystems.

In 1964 the Nobel prize for the development of the optical quantum generator (OQG) was jointly awarded to Drs. N.G. Bassov, A.M. Prokarov (members of the Soviet Academy of Sciences) and Charles H. Townes, an American scientist.

As is well known today, the early crystal lasers, and later gas and dye lasers, have been used in medicine to accomplish delicate surgery as "scalpels of light" to penetrate deep into the body where no knife can follow. In both eastern and western hospitals, this method is in practical use. We know also that this effect is primarily a thermal effect, and that the lasers involved must be of very high power to accomplish this task. By using pulsed, Q-switched lasers, it is possible to make a laser "drill" which, because of the high peak power, actually evaporates the tissue instead of burning through it. However, this also is a thermal effect and requires extremely high power densities.

These are certainly milestones in the medical use of lasers, however I think it ironic that in the race for higher power, we may have forgotten a basic question: are there any effects due to monochromatic light which can have medical usefulness, even when there is no thermal energy in the beam?

I have noted in the western literature a definite trend towards the use of higher and higher powered quantum generators, including the use of lasers to generate injection plasmas for fusion research. This follows the general trend in physics most evident in particle physics towards bigger and bigger systems, higher energies, larger and larger beam densities and to the huge accelerators which cannot even be operated by a single scientist. This type of research, while undoubtedly interesting, is enormously expensive, involves large research teams, and is based on the yet-unproven hope that there will be some significant resonances at these higher and higher energies. If these higher resonances do not exist, the cost in wasted man-hours and appropriations will be enormous.

In biological systems, the relevant energies are measured in terms of electron volts, not MEV; millijoules, not kilojoules. Is it possible that there are extremely low-energy resonance effects which fall into the near-optical waveband, which could be investigated by a laser using not kilowatts, but milliwatts of energy, and costing not billions of dollars but within the range of university department research budgets? Is it possible that very weak monochromatic, polarized light may have dramatic biological effects? The evidence which I have gathered from fellow researchers in the USSR and elsewhere indicates that both questions should be answered in the affirmative.

In October 1976 I was invited to take part in an unusual interdisciplinary seminar on the biological effects of laser light held at the department of biophysics at the Kazakh State University in Alma-Ata, capital of the Soviet republic of Kazakhstan. At this meeting I was able to meet many of the researchers in the field, and was allowed to tour various laboratories and a hospital where monochromatic polarized light is used in a variety of clinical and experimental settings. Since this is the first time that a physicist from a western university has visited these laboratories, this material is presented here to a wide audience for the first time.

Which Kind of Light Is Used In Biological Experiments?

The number of reports involving biological action of weak laser light is largest for visible red light, either from ruby lasers at 6943 Å or He-Ne at 6328 Å. This is partly because the red lasers were among the first to be available to biologists. More recently, UV gas lasers have been used. The most popular laser at the moment is the 35 mW He-Ne laser LG-75. Lasers which produce circularly or elliptically polarized light are preferred to those linearly polarized or unpolarized. The possible biological significance of this will be noted later.

Experiments have been in progress in this area since 1965, when research was started off on the effects of monochromatic red light (MRL) from conventional light sources, such as gas discharge tubes which were suitably filtered to make a pseudo-monochromatic source. As soon as lasers became available, these early sources were replaced with truly monochromatic sources.

Gas lasers, including helium-neon, carbon-dioxide, argon and xenon, have been developed to extend the range of wavelengths of radiation from 3000 Å to about 160,000 Å including the entire visible band.

Some 24 different semiconductor materials have also been used to achieve laser action, including GeAs, although many must be operated at low temperatures, some operate at room temperature, including a 1-watt laser at 9100 Å.

Several hundred dye compounds have also been found that will allow laser action. In 1973, V.I. Stepanov and A.N. Rubinov at the Byelorussian SSR Academy of Sciences constructed the "raduga" (rainbow) laser described in figure 3. A ruby laser 6943 Å is by means of a non-linear crystal converted to 3470 Å, which can be focused on dye cuvette chambers, which can hold up to 10 different dyes in a revolving drum, so the laser can be turned from 3600 Å to 10,000 Å, with a spectral line width of 50Å. In later models, the spectrum can be changed in only 10^{-5} sec.

Such experimentation has spread to other Soviet republics, and now the Tselinograd Medical Institute, the Lvov Medical Institute and the Moscow Medical Institute are also involved in similar work.

Psychical Self-Regulation

In 1973 the first all-union conference in the Soviet Union on the problems of "psychical self-regulation" (PSR) was held in Alma-Ata, the published proceedings being edited by Dr. A.S. Romen, a physiologist. For the first time, the concept of bioresonant stimulation was presented as being within the framework of psychical self-regulation, a process whereby biological control loops are activated by a variety of methods. By PSR the Soviets mean such diverse techniques as autogenic training, hatha yoga, and instrumentally augmented learned control of internal states biofeedback. These concepts, although known for some time, have only recently been seriously taken up by Western scientists. This conference proved such a success that a second thematic symposium with the same title was held in 1974. Neither of these conferences was open to Western scientists, although one U.S. scientist presented a paper in absentia at the second symposium.

In 1975 a book by Drs. V.M. Inyushin and C.R. Chekorov entitled "Biostimulation through laser radiation and bioplasma" appeared published by the Kazakh University Press with a circulation of 4,000 copies, the first textbook on MRL research. (For comparison, the previous conference volumes were only published in 1,000 copies.) I found this book of such interest, that I translated it into English and made it available to a small group of medical researchers who were interested in new medical therapy devices. This book summarizes the work carried out during the preceding nine years on MRL biostimulation in various parts of the USSR. The authors are esteemed colleagues of medical science: Dr. V.M. Inyushin is a professor of Biology at the Kazakh State University, and holds a state patent for the invention of the first "therapy box" using MRL. Dr. P.R. Chekorov is assistant minister of health for the Kazakh SSR and chief of a large hospital in the republic, and the first to use laser therapy in clinics.

In October 1976, a special seminar on "Problems of Bioenergetics of the Organism and Stimulation by Laser Light" was held at the Kazakh State University, chaired by Dr. Inyushin. For the first time, Western scientists were able to participate in discussions of the new medical treatment technique. Besides myself, attending were two Norwegian MD's, both trained in acupuncture, a Polish physical chemist, and an occupational therapist from Copenhagen.

The self-regulatory abilities of the body are really responsible for the many healings of medical science. No drug can cure an illness, unless the body's own mechanisms are functioning in the correct way. Therefore, we may consider that all illnesses, and all healings, are due to some aspect of the systems which control biological self-regulation.

Experiments in the biological effects of MRL have been ongoing since 1965 at the Kazakh State University named after S. N. Kirov. Besides the university labs, research is now conducted at the Alma-Ata State Medical Institute, the Aksai Republic Clinical Children's Hospital under the Kazakh ministry of health, the City Children's Hospital of Alma-Ata, and the Kazakh Scientific Research Institute for Tuberculosis. Elsewhere in the USSR I was informed that the experiments on MRL are being conducted in Moscow, Leningrad, Novosibirsk, Kazan, and Novgorod.

In 1967 the Kazakh State University Press published a booklet of conference proceedings on "Questions of Bioenergetics" which included topics on biological interactions of weak monochromatic light. It was only several years later when a private group in the U.S. translated this document and made it available to a limited audience. Conferences followed in 1969, 1970, 1971, and the number of scientific published papers in the field grew steadily in the USSR. The 1971 conference, which was titled "Problems of the bioenergetics of the organism in normal and pathological states" discussed for the first time the MRL work in the wider context of bioresonance stimulation.

It is well known that the damaging effect of laser radiation on biological tissues arises from the presence in the cell of components capable of absorbing light of a given wavelength. Thus, pigmented cells absorb the most visible energy, while unpigmented cells may absorb UV For example, mitochondria contain cytochrome-C, which has an absorption peak in the visible green.

In 1966, A.A. Gorodetskiy reported that red light from a ruby laser (6943Å) could bring about a free-radical state in irradiation of human blood. After irradiation, erythrocytes would give a stable EPR signal, indicating the presence of free radicals associated with the pigmented clusters. Human skin (containing the pigment melanin) also gives a stable EPR signal after ruby radiation.

The researchers in Kazakhstan had noted already in 1965 that the monochromatic polarized light (MPL) from a conventional source could have bioactive effects. The polarization of the red monochromatic light was altered by passing through layers of polarizers which gave elliptically or linearly polarized light. Researchers reported that there was a difference in effectiveness of the light action with identical wavelength, bandwidth, and intensity, but different only in polarization.

In 1973 Z.G. Beyasheva and B.A. Bekmuhambetova performed an experiment to compare the action of red polarized light, white polarized light, and HeNe laser light on man. The electroencephalogram (EEG) was monitored from the frontal and occipital lobes. When red or white polarized light was shone into the eye of the subject, the amplitude of the frontal EEG increased, while being simultaneously suppressed in the occipital EEG (which shows information processes in the visual cortex). HeNe laser irradiation, however, caused a total suppression of all EEG rhythms.

Perhaps the most dramatic biological effect of laser light is the stimulation of regeneration of tissue. In 1967 Dr. Koritniy and associates carried out experiments on rabbits using monochromatic polarized light from 6300-6400Å. Identical wounds were inflicted on both ears of a rabbit, and thus each rabbit served as its own control: one ear was irradiated, one was not. Distinct physiological changes including an increase in the number of neutrophyles occurred on the test wound but not the control wound. In successive periods, the number of phagocytes began to increase, and the bacterial flora composed of ciplo- and tetracocci almost entirely phagocytosized in the wound exudate. The wound which was irradiated with MRL also healed faster. The authors concluded that the effects were due to changing properties of the skin.

During the 1976 symposium in Alma-Ata, I asked Dr. Koritniy why he thought the number of bacteria decreased in the irradiated wound. Was the laser light's action bacteriocidal? No, he replied, their work showed that the laser light had no direct damaging effect on the bacteria, but due to the fact that the natural resistance mechanisms were improved, the bacteria were destroyed by increased phagocytal activity.

Koritniy has gone on to study the morphological and histological changes in laboratory-induced wounds caused by skin grafting. Microscopic changes were noted in the transplanted tissue. The leukocytal wall in the radiated skin shreds was considerably more prominent in comparison with those that were not radiated. An acceleration of inflammatory process phase changes was noted, and a stimulated proliferation of fibroblastic elements. In the unradiated sample, fibrose nutrients, which usually show up in the whole area, were also noted. This was noticeably absent in the radiated segments.

Simultaneously, in the surrounding connective tissue, the re-epidermization of the radiated transplant slows down. Near the 20th day after radiation, the whole transplant again becomes covered by epidermis. Other changes were noted in epithelia: in comparison with the control samples, the movement of epithelia along the connective tissue is slower, by about ten days. However, after the delayed phase, the new epithelial covering of the radiated transplant increases its productivity. In general, the specific organ structure of the radiated transplants re-establishes itself to the level which the unradiated samples took two or three months to achieve. The power of the laser radiation was only 0.2 to 0.5 mW/cm^2.

These and other experiments point to monochromatic polarized red light as being a powerful stimulating agent in regenerative processes. Other experiments have shown that regeneration of callous tissue in healing bone fractures is also stimulated by MRL radiation. Initially in a somewhat gruesome experiment, bone fractures were induced in dogs, and other controlled experiments (as mentioned above) were performed on rabbits. Today, human patients have also been treated for skin wounds such as lesions or trophic ulcers (unhealing sores), as well as broken bones with MRL radiation.

Experiments on Man: Regeneration of Tissue

Medical researcher K.D. Durmanov investigated five patients suffering from trophic boils in Alma-Ata, some of them had been suffering up to nine years, and all for at least one year. The boils were variously on the torso, the face, the knee, and the sole of the foot. MRL radiation of 25 mW/cm^2 was applied for 1.5 minutes. This dosage was given daily for 20 days. In all cases, therapeutic effects were noted, resulting in the total healing of the sores.

In 1972 V. V. Makeyeva of the Republic Clinical Hospital in Alma-Ata used a He-Ne laser in treatment of trophic sores in 25 patients. The wounds were a result of different causes, and some were as large as 27cm^2. All had previously undergone medical or operative treatments, but without success. They had been suffering from 12 months up to 25 years. The wounds were irradiated daily for 20-30 seconds. After 25 sessions, nineteen patients had completely healed wounds, and in four patients, the wounds had diminished in surface area. Already after the third to fifth session in the majority of the patients a considerable betterment of the wound condition was observed, and a disappearance of pain in the area of the wound and normalization of sleep patterns was reported.

A number of interesting case histories were presented at the conference. This is not the place to go into all the clinical details, but some background is needed to discuss possible mechanisms of interaction between light and biological systems.

The fundamental question is: can the action of laser rays be accounted for by a purely thermal hypothesis or by a hypothesis "of many factors," i.e. not only thermal.

In general we know that light has quite varied effects on life processes: for example red light has very little effect on photosynthesis, while blue and violet light can bring about some suppression of life activity. Ultraviolet light in the range of 2,500Å - 2,600Å, when absorbed by nucleic acids in cells, can be lethal. Since biological molecules (organic) have strong absorption peaks near the visible band, it may be possible for multi-photon processes to take place. It was noted, for example, that using a neodymium laser, lethal effects occurred at 10,600Å and 5,300Å. One absorbed quantum at 2,650Å is equal to the absorption of 2 photons at the second harmonic of 5,300Å or a 4-photon absorption at the original 10,500Å beam.

But how is the action of laser light different from that of conventional light sources of the same intensity and color? The optics of conventional sources and laser sources differs considerably. Even if the light intensity and center frequency of the emission is identical, the laser beam will have a narrower bandwidth, i.e. is more monochromatic, and will also be coherent in time and space. Semiconductor lasers are less spatially coherent, and the bandwidth is much broader. Strangely, semiconductor diode lasers do not seem to have any bioactive effect, while gas and crystal lasers do. This is an interesting point which I will come back to in the concluding section.

As to the biological mechanisms involved, considerable biochemical and biophysical studies have been made in the USSR. As an example let us take the work of A.I. Semonova and V.A. Singayevskiy which was published in 1969. They attempted to explain the biological action of low-intensity radiation by using 68 white rats. The biological reaction studied was the hormonal reaction of the suprarenal or adrenal glands, which in man are located bilaterally above the kidneys. Radiation was given from He-Ne gas, ruby crystal, and Nd++ glass lasers. He-Ne visible red radiation (6328Å) was applied continuously (CW) and brought about a reaction in the suprarenal glands within 10-20 minutes.

The changes noted involved a decrease in the number of iozynophiles in the peripheral blood by 20-40%, as well as a decrease in the content of lipids in the suprarenal cortex. These changes reached their maximum value 15-20 minutes after radiation was initiated. Two to three hours after radiation, the hormonal reaction decreased to its initial level.

In rats with a weakened nervous system, the return to the initial level took more than 5-6 hours. A similar hormonal reaction was not observed during or after irradiation with pulsed lasers, even after many exposures. Apart from the hormonal reactions, a sharp increase in the demand for hydrogen, from 50-100%, was observed. Seminov feels that a single gas laser radiation exposure is equivalent to many repeated radiations by pulsed ruby or Nd glass lasers.

One experiment with animals used a low-intensity laser ray which was directed into the eyes. The researchers noted changes of a functional nature in the cardiovascular system, such as changes in the tone of the vessels. After adaptation of the animals to darkness, the radiation brings about a marked displacement in arterial pressure, averaging about 20-30 mm Hg. Arterial pressure tends to decrease, while activity of hollyesterases increases sharply.

In rabbits, a shift in the ionic balance between calcium (Ca) and sodium (Na) was noticed in the reticular tissue of the eye. The authors theorize that the effect of coherent gas radiation is bioactive on the parasympathetic nervous system, i.e. a "vagotropic" effect.

Bioresonance and Cell Radiation

The idea of considering the body or part of it to be a resonating cavity is an interesting one. Just as we can have physical and chemical or electrical resonators we can also have bio-resonators. On the whole-body level, considering man as a somewhat wet leather bag filled with airspaces and liquids under pressure, we can expect that resonant waves of a system one meter in height would produce a wave on the order of meters or less. If we consider subsets of resonators in man, such as the airways, the fluid chambers, etc., as the magnitude of the system becomes smaller, the frequency proportionally increases. On the organ level, the wavelength of the signals would be in cm, at the cell level in mm, or sub-millimeter waves.

In this context we could be talking about acoustic waves, plane electromagnetic waves, or some kind of polarized EM wave, including visible light, IR and UV emission.

Let us restrict ourselves here to EM cellular emissions, which according to Soviet scientists, extended all the way from the radio band up to UV. The first evidence that such cellular radiations exist can be traced back to the Soviet histologist A.G. Gurvich (or Gurwitch) who in the early 1920s had noted in cell growth experiments that cells seemed to emit some kind of ray, which he termed "mitogenetic" or "mitotic" since it seemed related to cell reproduction by division, or mitosis. Although Gurvich had biological evidence that such a ray existed, he did not succeed in capturing it on film in his day. In fact, it was not until the early 1960s that workers at Leningrad State University succeeded in capturing the mitotic rays on sensitive photomultiplier devices.

It has been established by Zhuralev that the wavelength of the mitogenetic radiation lies between 190-350 nm, in the UV band. This radiation does not resemble other kinds of phosphorescence or luminescence known to biology.

It appears to be triggered at only 2 points in the life cycle of a cell, during birth (mitosis) and during death. In 1973 this aspect was verified by a team of Australian researchers who, using a giant tank of yeast, succeeded in counting UV emissions during growth and death of cells.

According to the Soviet researchers in Kazakhstan, the most intense UV mitogenetic radiation comes from the cell nucleus, which contains the recombinant DNA which is, of course, required for all cell division and growth.

In models of the cell being presently advanced by Popp, the cell DNA can play the role of an inductor at VHF frequencies, and due to its precise spiral pattern can maintain a definite phase relationship to the incoming waves. If the DNA-inductance is connected in parallel with the cell membrane capacitance, what results is a tank circuit capable of resonance at UV frequencies. The energy of the photons involved in this process is 3.6 eV, an energy particularly important for biological processes. Popp theorizes that there may be a two-level decay process from excited electronic states, giving rise to emission of coherent, polarized cell UV radiation – a biological laser.

According to the Soviet scientists, the cell radiation they have detected is both coherent and monochromatic polarized radiation: biolaser radiation. Wild as this may sound, it fits in with a number of data points which have emerged in recent years regarding the interaction of light and life.

In addition to this mitogenetic radiation, there are emissions from the cell and cell organelles at other frequencies, including the visible part of the spectrum. The mitogenetic radiation today is called in Soviet textbooks "dark chemoluminescence in the UV." It is interesting that this radiation is not mentioned in U.S. or U.K. biology texts.

According to the Soviet researchers, the cell mitochondria emit monochromatic red light from 6200Å-6400Å. This is, in fact, exactly half the frequency of the UV emission from the cell nuclei. The mitochondria are active in the process of cellular respiration, they mediate ion exchange and are storage points or reservoirs for stored cellular energy. The ion density inside the mitochondria, according to Polish sources, is higher than in the E-level ionospheric layer. Other parts of the cell emit radio frequency waves, according to Kazakh scientists.

This mitogenetic radiation is not the only kind of bioradiation from cells and organs. Soviet scientists have discovered several other wavebands of activity, which however are not related to the physiological "cold light" luminescence of lower animals which have a special fermentative apparatus.

In 1961 it was reported that mammalian cells showed endogenic luminescence and that lipid proteins also exhibited chemiluminescence. Further, this bioluminescence was divided into 2 groups: spontaneous and induced. Spontaneous superweak metabolic luminescence was reported in plants in 1954 by Italian workers Colli and Faccini, and in 1961 B. N. Tarusov and A. I. Zhuralev reported bioluminescence in mammals. According to Zhuralev, et al, this production of biophotons accompanies metabolic processes, primarily oxidizing reactions, and also destructive processes which disrupt metabolism: cancer, poisoning, overheating, etc. According to the Soviets, this is a universal mechanism, however since oxidation of free radicals in tissues is impeded, the intensity of this radiation is very low.

In 1964 Yu. A. Vladimirov discovered luminescence of cell mitochondria and these were later found to be tied to Fe^{++} ions. The waveband of superweak metabolic luminescence is 360-800 nm. Under normal conditions this cannot be detected by the naked eye. However, specially sensitive retinas may be able, under low-noise conditions, to detect this weak emanation from tissues.

Thus, biophoton emission could conceivably be used for remote diagnosis, and it is most likely that detectors or scanners of this sort are already under construction or in use on an experimental basis.

Induced Biochemiluminescence

In addition to the spontaneous radiation discussed above, the Soviet scientists note that stimulation from outside can induce this radiation in certain cases. Such triggering agents can be ultraviolet radiation, ultrasound, radio waves and ionizing radiation. Zhuralev et al have classified such induced radiation in 3 groups: (1) photosynthetic luminescence, (2) thermoluminescence, (3) photochemiluminescence, radiochemiluminescence and ultrasound luminescence.

The study of quantum phenomena in membranes, cell organelles and the cell nucleus may lead to the next breakthrough in quantum biology and biophysics. Interestingly enough, induced luminescence may have been discovered before endogenic radiation, at least as far as photographic imaging is concerned. As some of the other papers at this symposium will document, it is possible by irradiation of a sample by high-voltage, and moderate to high frequency-waves, to induce a corona discharge which contains information from several sources. In the most common configuration of corona-discharge photography (CDP) the light which is registered contains photons from gas emissions, metal emissions, and emissions of biophotons induced from the biological sample. In Kirlian photography (KP) this light is suitably filtered and processed to retrieve the biophoton information, which makes up an estimated 15% of the total emission. Kirlian scanners are being used on an experimental basis in the USSR for medical diagnosis. If it is indeed possible to characterize malignant tissue by the quantity or quality of biophotons it emits, then having a non-invasive scanner of this type would be an invaluable asset in medicine.

According to St. Györgi's [sic] theory of cancer, bioelectronic balance in the "sea" of free electrons leads to regulation of cell growth. Thus cancer is not a molecular disease, but an electronic disease. Any method which then can give quantitative information about the electronic balance (saturated or unsaturated) can be used in diagnosis as a parameter to observe during treatment. If bioelectronic measurements can be made painlessly and non-invasively, then it would be possible to monitor instantaneously the progress of any therapy.

The introduction of techniques which allow the measurement of bioelectronic parameters without surgical invasion or pain to the patient have been developed during the last decade and are now in experimental use in European labs. The technique of Vincent which uses measurements of Ph, redox potential (rH2) and electrical resistance (r) of the three major body fluids: blood, saliva, and urine, provides a 3x3 matrix of values which have been displayed in three dimensions to graphically plot the various bioelectronic areas of interest: the healthy norm, the area of infectious diseases, and zone of degenerative diseases. The graphs look not unlike the 3-D steam tables used by engineers to plot various thermodynamic variables which are not mutually independent.

Vincent claims that he has by this method been able to diagnose serious diseases such as cancer many months prior to the time when an X-ray, mechanical examination, or biopsy would reveal the growth. According to the theory, the bioelectronic balance of the patient changes prior to the actual development of the pathological nidus. The pH indicates the amount of free hydrogen ions in blood, saliva and urine and should these not be equal in the fluids, indicates that there may be an acid-base imbalance in the system. For example, should the blood pH be higher than the urinary pH, this indicates that the kidneys are not excreting acidic wastes properly. The rH2 parameter, which depends also on pH value, gives us a measure of the number of free electrons in the relevant fluid, something which has already been mentioned as being of importance in the electronic structure of biological material.

In investigations carried out in the U.S., measurements of electrical potential have been made on malignant vs. healthy tissue by Cone and co-workers. This may lead to an electronic screening method whereby abnormal cellular potentials can be scanned automatically in a manner analogous to the Vincent bioelectronic parameters. It would seem to the author that the measurement of crucial biopotentials would be an advantage to measuring the passive DC or AC resistance of cellular fluids. The bulk resistance rho (r/cm^3) is related to the sum total of electrolytes in solution, the charge carriers in the fluid being measured. Thus r gives us information not only of Na++ and Ca++ but also of all other elements, even those which are present only in trace quantities, such as Mg, Mn, Au, etc. Should these methods be properly evaluated, they may allow for an early warning approach to all disease processes, and regardless of the method of treatment, give a quick and detailed feedback of how the therapy process is proceeding. For example, suppose the therapist decides on a combination therapy which includes laser ray stimulation of certain acupoint areas and dietary changes to change the pH/rH balance. The points to be stimulated are chosen according to bioelectronic measurements of r and/or surface potential. Therapy is continued until the potential measurements are normalized, or until the bioelectronic plot shows a regression to the healthy zone. Any dietary change, such as increasing intake of pH-changing foods such as concentrated lemon juice, also shows up as a change in the 3-D plot.

This technique, although presently used in clinics in France, West Germany, and other European countries, is to my knowledge largely unknown in the USA.

Another type of bioelectric measurement which involves relatively quick and painless measurements is the diagnosis technique of electroacupuncture such as practiced in Japan, Europe, and the USSR. There are a number of different schools, each of which has developed its own method, but basically the measurement types can be classified as AC or DC impedance measurement or AC or DC potential measurements.

In the technique of Voll which is widespread in Europe, an indifferent metal electrode and a different metal electrode (usually the same metal) are applied to areas of the patient's skin and a small DC current is passed through the circuit. The different electrode is placed over an acupoint, and the indifferent electrode is placed at a remote location. The resistance is measured and displayed on a nonlinear scale arbitrarily divided from 0 to 100. In reality, 0 represents infinite resistance, i.e. open circuit measurement, while a reading of 100 means that there is a short circuit and maximum current is passed through the acupoint. These measurements are of course pressure-dependant, as the contact resistance between electrode and skin can cause the impedance to vary over a large range. Therefore the Voll method involves a substantial pressure on the skin (often somewhat painful) to reach a "plateau" region where the resistance value is relatively insensitive to small fluctuations in pressure. Occasionally, measurements are made with a calibrated pressure-sensitive probe, but often this is omitted. As the measurements are taken rather quickly, there is not electrode polarization, however, an abrupt fall (Zeigerabfall) or rise in the value (Zeigerabsteig) is given special significance, due to the "stability" of the points/meridians. The scale is arbitrarily centered at 50, which represents a resistance of approx. 30 k ohms. This is the "normal" range. Readings greater than 50 are taken to be indicative of infectious diseases (inflammatory conditions) and readings of less than 50 of degenerative diseases.

A rival school, founded by the late Dr. Croon, uses the same basic idea but instead of making DC impedance measurements, uses AC signals at several Khz. The impedance is resolved into resistive components and capacitive components which are then printed out on a "somatogram." In this measurement, no arbitrary center measurement is used. Instead, by visual inspection, the investigator assigns a median measurement which is then used to measure all deviations.

Since it is rarely that all meridians have abnormally high or abnormally low values, this technique allows each subject to serve as his own control, thereby making it unnecessary to assign a single median value to a large population with differing basal skin resistance. In practice, the Voll method to be accurate must vary the mean value of "50" depending on the actual value of skin resistance the patient has at the time of measurement.

Japanese Electroacupuncture

There are interesting similarities and differences between the German and Japanese schools of electroacupuncture. I will briefly describe 2 methods: the ryo-do-raku method, and Motoyama method.

In ryo-do-raku, a low-frequency AC impedance measurement is made. Only the resistive component is used in diagnosis. The use of a wet electrode of special construction allows for a method that uses very low contact pressures, only a few grams. Thus the method is less painful than the Voll method. A major difference is that meridian values are measured, but not acupoint values, i.e., the values are taken from specific points on the meridians which are not loci.

In the AMI technique used by Motoyama, a minicomputer is utilized to increase the speed and accuracy of the measurements. Instead of mechanically moving the search electrode from one acupoint to another (as in the Voll and Croon methods) the investigator can permanently attach all acupoints of interest (10 finger points, and 10 toe points, the so-called Seiketsu points, or terminal loci) and let the machine test for each value at high speed, store the data, perform statistical calculations on the evaluation of the data (which can be compared with case histories/norm values stored in memory) and print out a complete meridian diagnosis and suggested therapy regimen. The AMI technique used pulsed DC, 0-3 V positive, which are applied to the points for only a few ms. The transient response to the voltage pulse can be analyzed on an oscilloscope or in the minicomputer according to a fixed program. This transient curve gives information which relates not only to DC impedance, but also skin polarization potential, and dielectric constant, i.e. capacitance. It is claimed that this method is completely automatic and can be favorably compared to the semiautomatic or manual methods which take up to 100 times longer to carry out. The Voll method, completely manual, often takes over 1 hour for the 20 seiketsu points plus adjacent meridian points.

The semiautomatic Croon method, which has electronic storage and plotting capabilities, but only one search electrode which must be moved by hand, takes at least half the time of the Voll method. The AMI method, including time for placement of electrodes on all points and computing, does not exceed 10 minutes. It must be noted however, that this method uses only measurements on the terminal points. Specially designed "clothespin" electrodes provide constant pressure during measurement.

Other Noninvasive Scanners

Besides the bioelectronic and electroacupuncture methods, in recent years other devices have been developed which use secondary induced biophotons as a signal which can be electronically processed to give biological diagnostic information.

One technique, "Electronography," converts the biophotons to visible light frequencies via a fluorescent screen.

A technique under development by Inyushin, but not yet published, involves sensitive UV scanners which detect secondary biophotons which are kicked out by the free-electron bioplasma gas during short periods of radiation by monochromatic lasers. It is hoped that this technique can be developed to include whole-body scanning techniques similar to electronography.

The paper by Dobrin, et al. shows that background radiation emitted by living subjects in the visible plus UV range lies up to 4-5 times the noise level of the photomultiplier tube. Perhaps by irradiating the body with a monochromatic light source, the induced secondary biophotons will exceed the signal-to-noise level reported by Dobrin. However, before this type of scanner can be used to provide diagnostic information, it will be necessary to obtain spatial, as well as photon counting information which could be accomplished by either using a fixed subject and rotating scanner, or a rotating subject and fixed scanner. A 2-D plot could be made of any projected planar surface, and with the addition of a phi coordinate, a "catscan" type of tomography could be developed.

The Roots of Chromotherapy

There is some precedent that comes to us from medicine which supports the conclusion that weak irradiation by photons can have significant biological effects. I refer to the classical field of chromotherapy, or treatment by colors which flourished in this country and in Europe and India over a hundred years ago.

In the late 1890s Dr. Niels R. Finsen of the University of Copenhagen Medical School invented the "light bath" for treatment of certain skin diseases and also other illnesses using carbon-arc, UV, and other kinds of lamps. He was able to cure tuberculosis with green radiation, and made a number of important observations on skin pigmentation and light-susceptibility which are classics in the field. Today, the Finsen institute in Copenhagen is a specialized medical center for the study of cancer. Although chemotherapy has replaced the light bath for most ailments, some skin problems are still treated in this way.

In India, Dr. Dinshah Ghadiali published in 1878 his massive "Handbook of Chromotherapy" which contained the following associations for the colors with parts of the body:

RED: Sensory stimulant, liver energizer, irritant, vesicant, pustulant, rubefacient, caustic, haemoglobin builder.

ORANGE: Respiratory stimulant, parathyroid depressant, thyroid energizer, anti-spasmodic, galactagogue, antirachitic, emetic, carminative, stomachic, aromatic, lung builder.

VIOLET: Splenic stimulant, cardiac depressant, lymphatic depressant, motor depressant, leucocyte builder.

GREEN: Emotional stabilizer, pituitary stimulant, disinfectant, antiseptic, germicide, bactericide, muscle and tissue builder.

Now this list has some interesting correlations with the Soviet research on MPL and MRL we have just discussed. For example, the Soviets have noted that red light does regenerate haemoglobin building in erythrocytes. We see "red cells" because iron oxide in haemoglobin has an absorption at this frequency.

Similarly many elements have specific absorptions which fall in the visible range, including:

RED: cadmium, hydrogen, krypton, neon

ORANGE: aluminum, antimony, arsenic, boron, calcium copper, helium, selenium, silicon, xenon

GREEN: barium, chlorine, nitrogen, radium, tellurium, thallium

VIOLET: actinium, cobalt, gallium

PURPLE: bromine, europium, terbium

MAGENTA: irenium, lithium, potassium, rubidium, strontium.

It can be seen from this abbreviated list that nearly every major body salt or trace element is included. And, as mentioned previously, when we consider more complex resonances such as provided by atoms in combination and molecules, the list of absorptions grows quite large. Larger groupings, such as proteins and nucleic acids, have resonant absorptions in the ultraviolet.

It is interesting to note from Ghadiali's table that violet stimulates the growth of leucocytes, which we call "white cells." According to Ghadiali, leucocytes only appear white under the light microscope but are really "violet cells" or ultra-violet in color. The Soviet researchers have noted particular stimulatory effects on the "white" leucocytes and lymph system, both crucial to the body's immune system against diseases of all kinds. In the Soviet experiments, the red laser radiation was presumably converted via a nonlinear process into UV harmonic radiation which then stimulates the "white" cells.

Although around the turn of the century chromotherapy had made a foothold in this country, interest in it has faded. It is possible that persecution by the FDA invoking its regulatory power in the area of "quack" medical devices has made it difficult for MD's who were interested in chromotherapy to obtain the necessary equipment. Whatever the reason, it is certainly high time that the medical profession had a chance to experiment with non-lethal forms of radiation such as weak monochromatic light.

Conclusion

I have tried to show in this paper that the mysteries between biological life processes and bioelectronics are just beginning to be explored.

Experiments made with very low levels of light, from various sources, including visible MPL quantum sources, have shown a variety of effects which cannot presently be explained as thermal interactions. It has been postulated that different properties of the biofield allow energy exchange at the electronic-resonant level or higher using biophotons and biophonons within the free electron/exton gas permeating and surrounding living cells. The concept of Bioplasma, which is consistent with several of the experimental findings, must however for the time being be considered a theoretical construct which awaits experimental proof.

The basic findings of Soviet research into bioluminescence and induced bioluminescence have been reviewed, and the story of the mitogenetic rays has been updated. The concept of bioresonance and psychical self-regulation allow us to view bioelectronic regulatory mechanisms in a new light. The concept of nucleic acid as an electronic-resonant structure which may have the form of a 4-level laser which emits phase-coherent biophotons which can exchange energy with biophonons similarly to energy transfer in a crystal lattice. These biophotons have sharply quantized energy states, i.e., frequency and color.

The concept of noninvasive scanners is illustrated from the techniques of bioelectronic balance, and response to certain prescribed test signals, as in acupuncture. The possibility of developing scanners in the visible and near-visible range has been raised. The misinterpretations of the Kirlian effect are reviewed, and the difference between CDP and KP are outlined.

The author urges that interdisciplinary research groups be established to verify and expand on the striking findings which exist for using the biological field as a new tool in diagnosis, treatment, and basic research into the nature of life.

Light Therapy

Light Therapy: Norman E. Rosenthal, M.D.
Treatments of Psychiatric Disorders, Volume 3
American Psychiatric Association
Task Force on Treatments of Psychiatric Disorders
APA Press, 1989

Although it is only in the past five years that light treatment (phototherapy) for depression has attracted widespread interest, the concept of light as a therapeutic agent is by no means a new one. Even before the development of the drugs that are the mainstay of biologic treatments of depression and other psychiatric ailments, the quality of environmental light was considered to be important for an individual's welfare, along with proper nutrition, fresh air and sufficient rest. The use of light baths was commonplace in Europe in the early decades of this century, and books were devoted to the subject (Humphris 1924; Kovacs 1932). However, there are important distinctions between these light baths and the phototherapy that has been the focus of recent therapeutic attention. First, descriptions of these earlier light baths indicate that the body was bathed in light while the face was shielded from it; second, ultraviolet light was regarded as an important therapeutic ingredient; and finally, the proponents of this treatment claimed it was helpful for many different conditions. In contrast, phototherapy, as it has been used in the past few years, has involved exposing the eyes to light containing very little ultraviolet light in patients suffering from specific conditions.

Mood Disorders

Most of the research in phototherapy has been confined to patients suffering from seasonal affective disorder (SAD), a condition characterized by recurrent fall and winter depressions alternating with non-depressed periods in spring and summer

(Rosenthal et al, 1984). There have also been a few reports on phototherapy in non-seasonal depression (Kripke, 198, 1988; Kripke et al, 1986; Yerevanian et al, 1986), and some papers have suggested that light may also be beneficial in treating jet lag and other abnormalities of circadian rhythms (Daan and Lewy, 1984; Eastman, 1988; Joseph-Vanderpool et al, 1988; Lewy et al, 1983, 1984). By way of giving due historical credit, it should be noted that Esquirol, one of the earliest clinicians to describe a patient with a syndrome resembling SAD, effectively treated the patient by recommending that he winter in sunny Italy instead of Belgium (Esquirol, 1845). Later in the 19th century, an insightful ship's doctor observed that his crew became lethargic in the dark days of an arctic winter, and he treated their languor with bright artificial light (Jefferson, 1986).

Lewy et al (1980) made an important observation that ushered in the modern use of phototherapy – namely, that nocturnal human melatonin secretion could be suppressed by exposing subjects to bright artificial light, but not to light of ordinary indoor intensity. It has been shown that melatonin in animals is secreted nocturnally by the pineal gland in a circadian rhythm generated by the suprachiasmatic nuclei of the hypothalamus. Light rays impinging on the retina are converted into nerve impulses, which influence the secretion of melatonin by connections between the retina and the hypothalamus (Tamarkin et al, 1985). This demonstration that one physiologic effect of light in humans, transmitted presumably via the hypothalamus, has a threshold intensity far higher than that required for vision, suggested that there might be other effects of light on the brain that require high-intensity light.

The Syndrome of Seasonal Affective Disorder (SAD)

The cardinal criteria for SAD, as outlined by Rosenthal et al (1984), are: (1) a history of at least one episode of major depression, as defined by the Research Diagnostic Criteria (Spitzer et al, 1978); (2) recurrent fall-winter depressions, at least two of which occurred during successive years, separated by non-depressed periods in spring and summer; and (3) no other DSM-III-R (American Psychiatric Association, 1980) Axis I psychopathology. These criteria have been modified and included in DSM-III-R (American Psychiatric Association, 1987) as "Seasonal

Pattern," a descriptor that may qualify any recurrent mood disorder.

Defined by the above criteria, most patients with SAD have been women with an onset of the condition in the early twenties. Besides showing the usual affective and cognitive features of depression, their winters are also generally characterized by typical vegetative symptoms: overeating, oversleeping, carbohydrate craving, and weight gain; fatigue and social withdrawal are also prominent features. Since these symptoms are not suitably represented by the Hamilton Depression Scale (Hamilton, 1967), it has been suggested that supplementary items be added to this scale to measure adequately the extent of the symptoms in this condition (Rosenthal and Heffernan, 1985). Some researchers have found that most SAD patients report a history of Hypomania during spring and summer (Rosenthal et al. 1984), whereas others have observed a recurrent unipolar pattern, with euthymia in spring and summer (Yerevanian et al, 1986).

Many patients with SAD report a marked responsiveness to changes in climate, latitude and lighting conditions. Thus, many have noted improvement in depressive symptoms following moves to warmer, sunnier climates and latitudes closer to the equator. Similarly, patients may respond favorably to an improvement in the weather or to being moved from a windowless office to one with a window. Conversely, deterioration in mood and energy levels is often observed when the amount of environmental light is reduced.

Although the above description is typical for SAD, many variations are seen. For example, the symptom pattern may more closely resemble that seen in endogenous depressions. A variant of SAD has been reported to occur in children and adolescents; this variant appears to respond to phototherapy (Rosenthal et al, 1986). A milder version of SAD with prominent atypical vegetative features, but without the affective cognitive or affective symptoms has also been observed. This has been called atypical SAD (Rosenthal et al, 1985), seasonal energy syndrome (Mueller and Davies, 1986), or subsyndromal SAD (Kasper et al, in press; Rosenthal et al, 1985). A winter condition of disrupted sleep without affective symptoms has been reported to occur in Norway, where it has been termed Midwinter Insomnia and has been shown to improve after treatment with bright light in the morning (Lingjaerde et al, 1986).

Seasonal depressions may occur on a regular basis at other times of the year. For example, regularly occurring summer depression has been described by Wehr et al (1987) and by Boyce and Parker (1988). There have not as yet been any tests of the efficacy of phototherapy in such summer depression.

The Efficacy of Phototherapy in SAD

Following effective bright light treatment of a single patient with recurrent winter depressions (Lewy et al, 1982), several research groups have undertaken controlled treatment studies of phototherapy in patients with SAD. Researchers have sought both to establish the efficacy of the treatment and to find out which parameters are important for obtaining a treatment response. Most researchers have used crossover designs in which each patient has been exposed both to the treatment condition presumed to be active and to an alternate (control) treatment condition in a random-ordered fashion. Using such a design, phototherapy has been found to be effective in six studies by the NIMH group (Jacobsen et al, 1986, James et al, 1985, Rosenthal et al, 1984, 1985; Wehr et al, 1986a, 1986b), two studies by Lewy et al, in Oregon (1987a 1987b), two studies by Wirz-Justice et al in Switzerland (1986), and studies by Hellekson et al in Alaska (1986), Terman et al in New York City (1987), and Isaacs et al (1988) and Checkley et al (1986), both in London.

Since so many groups have by now replicated the basic finding that bright artificial light has antidepressant effects in SAD and there have not to date been any studies to the contrary, it has now been generally accepted that phototherapy is a viable treatment for the condition. However, those properties of phototherapy that are necessary or optimal for achieving an antidepressant effect continue to be a focus of interest and research (Rosenthal et al, in press). There is a consensus that intensity of light is important and that the bright light (2,500 lux) has been found to be superior to dimmer light (400 lux or less) in several studies (Checkley et al, 1986, Isaacs et al, 1988, Lewy et al, 1982, 1987b, Rosenthal et al, 1984, 1986, Terman et al, 1987) have suggested that even brighter light (10,000 lux) may be superior to the conventional 2,500 lux (Terman, in press), but the use of such bright light remains experimental at this time.

Optimal timing of light treatment has been somewhat more controversial. It appears as though two hours of light treatment in the morning may be more effective than two hours in the evening (Lewy et al, 1987a, in press; Rosenthal et al, in press; Terman et al, 1987; Terman et al, in press). However, other studies have shown that light treatment may be quite effective if administered only in the evening (Hellekson et al, 1986; James et al, 1985; Rosenthal et al, in press; Wehr et al, in press) or even during the day (Isaac et al, 1988; Jacobsen et al, in press; Wehr et al, 1986a). A cross-center analysis supports the findings that morning light treatments are superior to evening treatments, though certain individuals respond to the latter (Terman, in press). Although early studies used five to six hours of light per day, more recent studies suggest that two hours of treatment per day may be effective (Lewy et al, 1987b; Terman et al, 1987; Wirz-Justice et al, 1986). In some cases, as little as 30 minutes of light in the morning have been reported to be effective (Lewy et al, 1987b). However, there does seem to be some relationship between duration and efficacy, with greater effects occurring as duration is increased (Terman et al, 1987; Wirz-Justice et al, 1986).

Although most studies have used full-spectrum light, it is unclear exactly what spectral properties are necessary to achieve an antidepressant effect and whether full-spectrum lighting provides any advantages. Only one study has thus far addressed the necessary route of administration of effective light therapy, which appears to be the eye rather than the skin (Wehr et al 1986b).

Practical Considerations in Phototherapy

We have not yet been using phototherapy for long enough to outline definitively the best way of administering it. The method outlined below represents the combined experience of several different centers.

The light source most frequently used has been full-spectrum fluorescent light (Vitalite). However, other, less expensive full-spectrum fluorescent lamps are available. Six Powertwist or eight regular 40-watt tubes are inserted into a rectangular metal fixture, approximately two by four feet, with a reflecting surface behind them and a plastic diffusing screen in front. Smaller, more convenient fixtures are currently being marketed, but studies comparing the efficacy of different fixtures are lacking. Patients are asked to place the box at eye level, either horizontally on a desk or table or vertically on the floor, and to sit approximately three feet away from it in such a way as to expose their eyes to the lights. The intensity resulting from this setup is 2,500 lux, the amount of light to which one would be exposed by looking out of a window on a spring day in the northeastern United States. This is five to ten times brighter than ordinary room lighting. We do not know whether it is necessary for patients to glance directly at the lights intermittently, as they have usually been asked to do, since we do not know where the relevant photoreceptors are in the retina. Indeed, only recently has it been shown that the effects of phototherapy are probably mediated by the eye and not the skin in most cases (Wehr et al, 1986b). If the relevant photoreceptors are the rods on the periphery of the retina, it may be unnecessary for patients to glance at the lights directly at all. It is certainly unnecessary, and probably inadvisable, for patients to stare continuously at the lights.

In treating a patient with SAD, the initial suggestion for timing and duration of treatment is a matter of clinical judgment; since it does not appear that timing is crucial for obtaining an effect in many patients, some clinicians initially recommend times that are convenient for the patient. An alternative approach would be to begin with light treatments in the morning, which produces optimal effects in many patients. An initial dose of two to four hours per day seems reasonable. In fact, there is some evidence that an extra two hours of treatment in the evening does not enhance the response of two-hour treatment sessions in the morning. The clinician should feel free to titrate the dosage up or down, depending on the patient's response after one week of treatment. When a response occurs, it is almost always apparent within the first four days (Rosenthal et al, 1985). If no response occurs within the first week, one can either increase the duration or alter the timing. It is critical at this point to check that the patient is complying with all aspects of treatment. Treatment may be given in conjunction with antidepressant medications, an approach that may enable the patient to be treated with a lower dosage of medication or may convert a partial light responder into a complete responder.

The technology of phototherapy is in its infancy. Although most researchers have used light fixtures similar to that described above, Yerevanian et al (1986) have described clinical success in nine SAD patients treated with indirect incandescent light exposure. As noted above, Terman et al (in press) have reported superior results with fixtures delivering higher lighting intensities. Other lighting delivery systems are currently being developed in an attempt to make treatment more effective and convenient.

Side Effects of Phototherapy

Although side effects are uncommon, patients sometimes complain of irritability (of the kind seen in hypomania), eyestrain, headaches, or insomnia. The latter is most likely to occur when patients use lights late at night. Side effects can generally be reversed easily by decreasing the duration of treatments or suggesting that patients sit further from the light source. In a few cases, however, treatment may have to be discontinued altogether because of severe eye irritation. Although hypomanic responses have been observed in several cases, very few cases of florid mania following phototherapy in any typical SAD patients have been noted to date. Thus far, there have been no reports of any long-term adverse effects of phototherapy when properly administered.

Mechanism of Action of Phototherapy in SAD

The mechanism of action of the antidepressant effects of SAD is not known. Rosenthal et al (1986) suggested that bright light might exert its antidepressant effects by the suppression of melatonin secretion. However, they went on to show in a series of studies that this mechanism is probably not of central importance in most cases. Lewy et al (1987a) have suggested that light exerts its antidepressant effects by means of its circadian phase-shifting properties, a theory that continues to be actively explored. It appears as though certain individuals have a neurochemical vulnerability (perhaps genetically determined), which, in the absence of adequate environmental light exposure, produces the behavioral changes seen in SAD. Bright light, probably acting via the eye and presumably via retino-hypothalamic projections, appears capable of reversing this biochemical abnormality if the light is of high enough intensity and is used regularly and for sufficient duration.

It should be noted that bright light has been shown to be capable of producing changes in the P300 component of the visual event-related potential (Duncan et al, unpublished manuscript), in the increase in plasma norepinephrine seen when a patient stands up (Skwerer et al, 1987), and in peripheral lymphocyte function, as measured by in vitro exposure of patients' lymphocytes to a variety of mitogens (Skwerer et al, 1987). Whether these observed biologic changes have any relationship to the antidepressant effects of light remains to be seen.

Treatment of Nonseasonal Affective Disorder

Studies on the effects of phototherapy in nonseasonal depressives have been conducted by Kripke (1985, 1988), Kripke et al, (1986) and Yerevanian et al (1986). Although recent data by Kripke et al (1986) appear encouraging the efficacy of phototherapy in this group is far less clear than it is in patients with SAD, and further studies are required before bright light can be recommended clinically for nonseasonal depressives.

Therapeutic Use of the Phase-Shifting Effects of Light

The timing of environmental light and dark is well known to be an important circadian time cue (zeitgeber) in many animal species (Rusak and Boulos, 1981). Wever and Aschoff, however, in their pioneering studies of humans in environmentally isolated condition, concluded that light cues are weak zeitgebers in humans (Wever, 1979). Czeisler et al (1981) subsequently showed that light and dark cycles were sufficient to entrain environmentally isolated humans, and Lewy et al (1984) suggested that bright artificial light might indeed be a more powerful zeitgeber than the light of lower intensity used by Wever in his earlier studies. Wever later used bright artificial light in free-running human subjects and found that it was indeed a far more powerful zeitgeber than his earlier studies had led him to believe (Wever, 1979). Since then, Czeisler et al (1986) have shown bright artificial light capable of shifting circadian rhythms in a single subject. Lewy et al (1987a) have shown that bright light is capable of shifting the timing of circadian rhythms, as measured by 24-hour temperature minima and the timing of onset of melatonin secretion when subjects are kept in dim light. They have suggested that this shifting function may be therapeutically beneficial in depression and disorders of circadian timekeeping (Lewy et al,

1984). In a study of one patient with a temporary disorganization of circadian functioning, namely jet lag, Lewy and Daan (1984) showed that judiciously-timed bright light exposure appeared to ameliorate the jet lag symptoms.

In animals, light stimuli are capable of shifting the timing of circadian rhythms either earlier or later to different degrees, depending on when the light stimulus occurs in the course of the animal's daily rhythms. The relationship between the timing of the light stimulus and the resulting direction and extent of shift of circadian rhythms is known as the animal's phase response curve to light (Pittendrigh, 1981). This characteristic biologic functioning has been observed in many different species and appears to conform to certain rules. In general, light stimuli occurring tend to shift the animal's rhythms earlier. Conversely, light stimuli occurring early in the animal's subjective night tend to shift its rhythms later. These basic principles would need to be borne in mind in evaluating the effects of bright light on human circadian rhythms.

Preliminary results from a study of patients with delayed sleep phase syndrome suggest bright light in the morning, combined with dim light in the evening.

The Therapeutic Value of Light and Color

Abstract of a paper presented at the clinical meeting of the Section on Eye, Ear, Nose, and Throat Diseases of the Medical Society of the State of Pennsylvania, held at the Medico-Chirurgical Hospital, Philadelphia, October 12, 1926.
— *Atlantic Medical Journal*, April 1927

For centuries scientists have devoted untiring effort to discover means for the relief or cure of human ills and restoration of the normal functions. Yet, in neglected light and color there is a potency far beyond that of drugs and serums.

In the effort to obtain relief from suffering, many of the more simple but potent measures have been overlooked while we have grasped at the obscure and complicated.

Sunlight is the basic source of all life and energy upon earth. Deprive plant or animal life of light, and it soon shows the lack and ceases to develop. Place a seed in the very best of soil or a human being in a palace, shut out the light, and what happens? Without food (in the usual sense of the term) man can live many days; without liquids a much shorter time; but not at all without the atmosphere which surrounds him at all times and to which he pays so little attention. The forces on which life mostly depends are placed nearly or quite beyond personal control.

In order that the whole body may function perfectly, each organ must be a hundred percent perfect. When the spleen, the liver, or any other organ falls below normal, it simply means that the body laboratories have not provided the required materials with which to work, either because they are not functioning, as a result of some disorder of the internal mechanism, or because they have not been provided with the necessary materials. Before the body can appropriate the required elements, they must be separated from the waste matter. Each element gives off a characteristic color wave. The prevailing color wave of hydrogen is red, and that of oxygen is blue, and each element in turn gives off its own special color wave. Sunlight, as it is received by the body, is split into the prismatic colors and their combinations, as white light is split by passage through a prism. Everything on the red side of the spectrum is more or less stimulating, while the blue is sedative. There are many shades of each color, and each is produced by a little different wave length. Just as sound waves are tuned to each other and produce harmony or discord, so color waves may be tuned, and only so can they be depended on always to produce the same results.

If one requires a dose of castor oil, he does not go to a drug store and request a little portion from each bottle on the shelves. I see no virtue, then, in the use of the whole white light as a therapeutic measure when the different colors can give what is required without taxing the body to rid itself of that for which it has no use, and which may do more or less harm. If the body is sick it should be restored with the least possible effort. There is no more accurate or easier way than by giving the color representing the lacking elements, and the body will, through its radioactive forces, appropriate them and so restore the normal balance. Color is the simplest and most accurate therapeutic measure yet developed.

For about six years I have given close attention to the action of colors in restoring the body functions, and I am perfectly honest in saying that, after nearly 37 years of active hospital and private practice in medicine and surgery, I can produce quicker and more accurate results with colors than with any or all other methods combined – and with less strain on the patient. In many cases, the functions have been restored after the classical remedies have failed. Of course, surgery is necessary in some cases, but results will be quicker and better if color is used before and after operation. Sprains, bruises and traumata of all sorts respond to color as to no other treatment. Septic conditions yield, regardless of the specific organism. Cardiac lesions, asthma, hay fever, pneumonia, inflammatory conditions of the eyes, corneal ulcers, glaucoma, and cataracts are relieved by the treatment.

The treatment of carbuncles with color is easy compared to the classical methods. One woman with a carbuncle involving the back of the neck from mastoid to mastoid, and from occipital ridge to the first dorsal vertebra, came under color therapy after ten days of the very best of attention. From the first day of color application, no opiates, not even sedatives, were required. This patient was saved much suffering, and she has little scar.

The use of color in the treatment of burns is well worth investigation by every member of the professions. In such cases the burning sensation caused by the destructive forces may be counteracted in from twenty to thirty minutes, and it does not return. True burns are caused by the destructive action of the red side of the spectrum, hydrogen predominating. Apply oxygen by the use of the blue side of the spectrum, and much will be done to relieve the nervous strain, the healing processes are rapid, and the resulting tissues soft and flexible.

In a very extensive burn in a child of eight years of age there was almost complete suppression of urine for more than 48 hours, with a temperature of 105 to 106 degrees. Fluids were forced to no effect, and a more hopeless case is seldom seen. Scarlet was applied just over the kidneys at a distance of eighteen inches for twenty minutes, all other areas being covered. Two hours after, the child voided eight ounces of urine.

In some unusual and extreme cases that had not responded to other treatment, normal functioning has been restored by color therapy. At present, therefore, I do not feel justified in refusing any case without a trial. Even in cases where death is inevitable, much comfort may be secured.

There is no question that light and color are important therapeutic media, and that their adoption will be of advantage to both the profession and the people.

Kate W. Baldwin, M.D.

Influence of Fluorescent Light on Hyperactivity and Learning Disabilities

John N. Ott, ScD, from the files of ELC

During the first five months of 1973, a pilot project was conducted by the Environmental Health and Light Research Institute in four first-grade windowless classrooms of a school in Sarasota, Florida. In two of the rooms, the standard cool-white fluorescent tubes and fixtures with solid plastic diffusers remained unchanged. In the other two rooms, the cool-white fluorescent tubes and fixtures were replaced with full-spectrum fluorescent tubes that more closely duplicate natural daylight. Lead foil shields were wrapped around each end of the tubes where the cathodes are located. Aluminum "egg crate" diffusers with an additional grounded aluminum screen grid replaced the solid plastic diffusers in these latter rooms. A dramatic improvement in behavior was demonstrated in hyperactive children.

* * * * *

Exploring the effect of lighting on behavioral problems is the newest emphasis for time-lapse photography at the Environmental Health and Light Research Institute. This is a report of a pilot project conducted in windowless elementary school classrooms by the Institute to study the effect of fluorescent lights on the behavior of children.

Method

Subjects for this study were children in four first grade, windowless classrooms in Sarasota, Florida.

In two of the rooms the standard cool-white fluorescent tubes and fixtures with solid plastic diffusers remained unchanged.

In the other two rooms, the cool-white fluorescent tubes were replaced with full-spectrum fluorescent tubes that more closely duplicate natural daylight. Lead foil shields were wrapped around the cathode ends of the tubes to stop suspected soft X-rays.

Procedure I: Special cameras were mounted near the ceiling in each of the four classrooms out of view of the children. The cameras were set to photograph sequences of time-lapse pictures during the school day.

Results: The photographs revealed the following: In the classrooms with standard unshielded fluorescent light children could be observed fidgeting to an extreme degree, leaping from their seats, flailing their arms, and paying little attention to their teachers.

In the experimental classrooms the first graders settled down more quickly and paid more attention to their teachers. Less nervousness was evident and overall performance was better.

Procedure II: Full-spectrum shielded lighting was then installed in the two classrooms with standard lighting used in the earlier part of this study. Two and three months later the same children were photographed in the classroom in similar time-lapse pictures.

Results: The photographs revealed a very significant difference in the behavior of these children. They appeared calmer and more interested in their work. One little boy who stood out in the earlier films because he was constantly in motion and was inattentive had changed to a quieter child, able to sit still and concentrate on classroom routine. His teacher reported that he was capable of doing independent study now and that he had even learned to read during the short period of time.

Discussion

The results of this study may indicate that hyperactivity is partly due to a radiation stress condition. Improvement in the children's behavior occurred when we eliminated excessive radiation and supplied that part of the visible spectrum which is lacking in standard artificial light sources.

Drugs and Hyperactivity: No drugs were administered in this study, and this is of particular significance since warnings are now being heard about the widespread use of amphetamines and other psychoactive drugs on children thought to be hyperactive. As child psychiatrist Mark Stewart of the University of Iowa pointed out in *Time* (Feb. 26, 1973), the danger is that "by the time a child on drugs reaches puberty, he does not know what his undrugged personality is."

Estimates of the number of children in this country now taking drugs range as high as 1,000,000, a situation which prompted the Committee on Drugs of the American Academy of Pediatrics to propose regulations to the U.S. Food and Drug Administration to prevent abuses (see *Time*, Feb. 26, 1973). Psychoactive drugs have been shown helpful in treating hyperkinesis, a restlessness that some experts believe derives from minimal brain damage or chemical imbalances. I worry for the future of the hyperactive boy we photographed and the many other children like him. If he gets relief through drugs from stress caused by malillumination and from radiation, will that lead to later addiction to drugs or alcohol?

Irving Geller, chairman of the Department of Experimental Pharmacology at southwest Foundation for Research and Education in San Antonio, has found that abnormal conditions of light and darkness can affect the pineal gland, one of the master glands of the endocrine system. Experimenting with rats, Geller (1971) discovered that rats under stress preferred water to alcohol until left in continuous darkness over weekends. Then they went on alcoholic binges.

Nobel Prize winner Julian Axilrod (1974) earlier found that the pineal gland produces more of the enzyme melatonin during dark periods. Injections of melatonin in rats on a regular light-dark cycle turned these rats into alcoholics.

Alcohol: I have found that many biological responses are to narrow bands of wavelengths within the total light spectrum. If these are missing in an artificial light source, the biological receptor responds as in total darkness. That alcoholism may be related to the pineal gland is also under study by Kenneth Blum, a pharmacologist at the University of Texas Medical School (see *Science News*, April 17, 1973). Under near total darkness, rats with pineals drank more alcohol than water while rats without pineals drank more water than alcohol. When the animals were returned to equal periods of light and dark, rats with pineals retained their liking for alcohol. Applied to humans, Dr. Blum says "it is possible that alcoholics may have highly active pineals."

Artificial Food Flavors and Colorings: The hyperactive reaction to radiation from unshielded fluorescent tubes may have a correlation to the hyperactivity symptoms and severe learning disorders triggered by artificial food flavors and colorings (see *Newsweek*, July 9, 1973). Ben F. Feingold of the Kaiser Permanente Medical Center in San Francisco found that a diet eliminating all foods containing artificial flavors and colors brought about a dramatic improvement in 15 of 25 hyperactive school children studied. Any infraction of the diet led within a matter of hours to a return of the hyperkinetic behavior (1973, 1975).

This points out the possibility of an interaction between wavelength absorption bands of these synthetic color pigments and the energy peaks caused by mercury vapor lines in fluorescent tubes. This could explain the reaction, or "allergy" to fluorescent lighting.

This could be eliminated two ways: by eliminating the absorbing material consumed when the child eats artificial coloring or by eliminating the culprit in fluorescent tubes. The fluorescent cathode as a source of soft X-ray has been recognized by such scientists as K.G. Emelius, professor of physics at Queen's University, Belfast, in his book, *The Conduction of Electricity Through Gases*.

I first suspected soft X-ray of having a deleterious effect on children as a result of time-lapse pictures that I made of flowers for the Barbra Streisand film "On a Clear Day You Can See Forever." Flowers nurtured under high-power fluorescent lights didn't grow as well near the ends of the tubes. Additional tests on bean sprouts showed abnormal growth in those near the ends of fluorescent tubes. With TV X-ray measuring equipment, I have detected slight measurements of X-rays at the ends of the tubes which would penetrate aluminum foil, but not lead foil.

The reaction to light energy through the eyes affecting the glandular system is being studied in another major research project being carried on by the Environmental Health and Light Research Institute. Anthony Marchese, working in the Department of Pharmacology at Stritch Medical School of Loyola University in Chicago, is studying the mechanisms in the eye that responds to the pineal and pituitary glands which control basic body chemistry through production and release of hormones. Marchese has completed four years of medical school and is working on this project for his PhD dissertation.

The non-visual response of the eye to wavelengths of light is why I have been so interested in the effects of tinted lenses, sunglasses and artificial light sources that produce a distorted light spectrum entering the eye. By chance I observed that my arthritis improved when I lost my glasses and had to work in natural sunlight without them. I found that ordinary window glass and automobile windshields shut out up to 90 percent of the ultraviolet rays. I helped develop the full-spectrum plastic lenses that are now available in clear and neutral gray. Contact lenses can also be ultraviolet-transmitting.

Radiation from TV Sets: My concern with the harmful effect of radiation from TV sets was tested in another school study, this time among hyperactive children placed at the Adjustive Education Center in Sarasota. TV sets in these children's homes which were found to be giving off measurable amounts of X radiation were either repaired or discarded. Sets were moved so that none would back up against a wall where anyone might be working or sleeping in the next room. Parents cooperated in making their children sit back as far as possible and restricting the number of hours the children could watch TV.

The school principal reported that an improvement had been noted in the behavioral problems of the children in whose homes TV sets had been found which had been giving off radiation and which were repaired or removed. She noted that one of the most hyperactive children had been sleeping on the other side of a wall from a TV set giving off .3 milliroentgens of radiation per hour through the wall. This is within the "safety" standards of .5 mrh set up by the 1968 Radiation Control Act. Without the TV radiation, this child improved so markedly she could return to her regular school.

Earlier research (Ott 1968 a.b.) showed that young rats placed close to a color TV set with the picture tube covered with black photographic paper became highly stimulated, then progressively lethargic. All died in 10 to 12 days. Other recent experiments showed abnormal biological responses in plants left close to the ends of fluorescent tubes. The cathode there is basically the same as in a TV picture tube or X-ray machine.

Conclusions

Under improved lighting conditions, using full-spectrum fluorescent tubes with lead foil shields over the cathode ends of the fluorescent tubes to stop soft X-rays, children's behavior in the classroom showed dramatic improvement. On the basis of these findings, the Sarasota School Board has authorized an expanded comprehensive study. The proposed expanded study in nine windowless classrooms, kindergarten through second grade, will compare behavior patterns, scholastic achievement, attention spans, and general health under the improved lighting environment with present standard classroom lighting.[1]

---

1 Environmental Health and Light Research Institute, 1873 Hillview Street, Sarasota, Florida 33579.

The Cases of
E.K. Knott

As mentioned in Chapter 2, Mr. E.K. Knott presented several cases before a meeting of the King County Medical Society in Seattle, Washington in 1944. These cases were treated and documented at Hahnemann Hospital in Philadelphia, and are summarized as follows:

T.G., a fifty-six-year-old white male, was admitted to the Hahnemann Hospital complaining of pain in both feet. Examination of his feet revealed a dry gangrene of the right great toe and a swollen tender area on the medial side of his left heel. Both ankles were swollen. On May 14, 1941, neither tibial nor dorsalis pedis pulses were palpable. The patient gave a history of having had his feet frozen several times in the last ten years. During the past summer the patient apparently had symptoms of intermittent claudication. In November, the patient noticed that the right great toe was sore and this continued until April, 1941, when he injured this toe and it became infected. The toe was incised to establish drainage but the toe only became worse and turned black.

He came to the accident ward seeking help and was admitted on May 12, 1941. The urinalysis, blood count, blood sugar, and blood urea were all normal. Wassermann and Kahn tests were normal. Blood culture was sterile after 120 hours' incubation. Temperature was relatively normal except for a small rise in temperature to 99°F each evening. Treatment was instituted as follows: five percent saline soaks were given daily for three days and then three times weekly. On May 16, 1941, the vacuum pressure boots were started on both legs. The saline and the boot seemed to aggravate the condition so that they were stopped on

May 22, 1941, and the patient was advised to have an amputation of the right leg above the knee because of the mortification setting in half-way to the right knee. On May 25th under cyclopropane-oxygen anesthesia the right extremity was amputated just above the knee using the guillotine method. Good hemostasis was obtained, the flaps approximated and the incision closed with drainage.

Following this operation the patient was very much improved and had very little discomfort. The sutures and drains were removed May 31st and the patient was ordered out of bed. Dressings were changed daily.

On June 12th, the patient began to complain of the left heel which had apparently been healed. Whirlpool bath was ordered daily. The patient then began having pain in the stump as well as in the left heel. On June 15th, a nerve block was ordered. Since the whirlpool bath apparently aggravated the pain, it was discontinued.

On June 17th, a lumbar sympathetic nerve block was done. Just before the block an abnormally high skin temperature was recorded. The patient was not relieved of pain and the temperature of the leg did not change. This was repeated on the 19th without relief to the patient. The patient complained almost constantly of pain in the left heel which had now become swollen and edematous. Pain continued until blood irradiation was performed June 28th. Following this irradiation the swelling went down in the left foot and the area on the left heel began to heal and granulate. The stump of the right leg was almost completely healed following treatment and the drainage was negligible.

On July 3rd the patient was discharged free of pain both in the right amputation stump and in the left heel, there was slight drainage from the left heel and the stump of the right leg.

* * * * *

E.J., a twenty-two-year-old black female, was admitted to the hospital on August 20, 1941. The diagnosis was Streptococcus hemolyticus septicemia and generalized peritonitis secondary to ruptured tubo-ovarian abscess. On admission she was complaining of severe abdominal pain and gave a history of sudden onset. Physical examination revealed a markedly tender, rigid abdomen characteristic of an acute intra-abdominal condition. Her pulse was 110, respirations 30, temperature 101°F. Pelvic examination revealed presence of fluid in the cul-de-sac. Laboratory examinations revealed negative urinalysis, hemoglobin 14.3 Gm., red cells 4,700,000, white cells 12,300. A diagnosis of ruptured ectopic pregnancy was made and laparotomy was performed soon after admission.

At operation generalized peritonitis was found which was secondary to a ruptured tube-ovarian abscess. The abdomen was closed with adequate drainage. The patient's temperature, pulse and respirations rose to 105°F., 140, and 36 respectively. Sulfathiazole was started the second postoperative day, August 22nd; the patient became extremely nauseated and started vomiting; her condition was grave. Sulfathiazole was stopped. Twelve hours later ultraviolet blood irradiation therapy was instituted, August 23rd, 1941. At this time blood culture taken on the day of admission was found to be positive, showing a pure and luxuriant growth of Streptococcus hemolyticus. The next forty-eight hours the patient seemed generally improved, temperature, pulse, respirations falling to 99°, 102, and 30, respectively.

On August 26th, the abdominal wound began to break down and the following day, because of a rise in respiratory rate to 38, ultraviolet blood irradiation therapy was repeated. The following day the temperature continued to rise as did the pulse and respiration, to 103.2°F., 148, and 46, respectively. On August 29th, a third blood irradiation was administered.

On August 29th, X-ray examination of the chest failed to show any definite pulmonary disorder but the patient was put in an oxygen tent because of marked dyspnea and the presence of rales throughout the chest. The patient's condition remained stationary and on September 2nd ultraviolet blood irradiation therapy was given for the fourth time. The following day her temperature rose to 105°F. Forty-eight hours later it began to fall. At this time, September 4th, it was believed that the oxygen tent should be removed because of the apparent deleterious effects of mechanical irritation and overventilation. On removal of the oxygen tent respiratory rate fell to a level between 24 and 30. Shortly after this a large, subcutaneous and subfascial abscess ruptured. On September 6th the patient was definitely out of danger and continued to convalesce uneventfully till October 4th when her temperature rose to 101.2°F. There was a daily temperature rise to 101° or 102°F. for three days following, during which time an inflammatory area appeared in the lateral region of the left buttock. On October 17th this ruptured and her temperature fell to normal. The patient convalesced uneventfully from this point on and was discharged in good condition on November 2nd.

* * * * *

A.W., a twenty-year-old white female, diagnosed as having puerperal sepsis and septic endometritis complicated by acute pyelitis, was admitted to the obstetrical ward November 24, 1941, at full term. Midforceps delivery was performed; third degree lacerations of the peritoneum occurred. The first four days following delivery the patient ran a moderately septic course, temperature ranging from 100° to 103°F., pulse 138 and respiration 34, respectively; severe chills were present; urinalysis showed 30 to 40 pus cells per h.P.F. The chills continued for 48 hours; the patient became nauseated, delirious, and increasingly toxic. On the seventh postpartum day, ultraviolet blood irradiation therapy was instituted. The following day no chills occurred, nausea was absent, the patient was mentally alert, and her temperature, pulse and respiration fell to normal levels; her toxic symptoms had generally subsided. This improvement was maintained with the result that she convalesced uneventfully, leaving the hospital on the seventeenth postpartum day, in apparently excellent condition, ten days after a single irradiation.

* * * * *

S.C., a white woman of fifty-two years, was admitted to Hahnemann Hospital on August 11, 1941. The final diagnosis was incomplete septic abortion complicated by lobar pneumonia. When admitted she was complaining of chills, fever, and abdominal bleeding, giving a history of approximately ten days of chills and fevers. Physical examination revealed a markedly toxic patient with a temperature of 103°F., pulse 120, respiratory rate of 20. Pelvic examination revealed a lacerated external os and vaginal bleeding. Laboratory examination revealed hemoglobin 11.2 Gm., red cell count 4,000,000, white count, 12,250, sedimentation rate 7 mm. in fifteen minutes, 30 mm. in forty-five minutes. Culture of the cervix showed the presence of Staphylococcus aureus and Bacillus diptherodeus. Sulfathiazole was started immediately on admission. Within a few hours her respiratory rate rose to 40 and some cyanosis was present.

The following day an X-ray examination of the chest showed the presence of consolidation characteristic of early lobar pneumonia. The patient's general condition continued to deteriorate, her dyspnea and cyanosis increased mental confusion, nausea and vomiting appeared, and her temperature, pulse, and respiratory rates remained elevated. Therefore, because after 72 hours of sulfathiazole therapy it was believed that the patient's condition had been in no way improved by sulfathiazole, and that, on the contrary, it had become very critical, sulfathiazole was discontinued.

Several hours later ultraviolet blood irradiation therapy was instituted; within a few minutes the patient's dusky cyanosis began to disappear and in its place was seen a definitely pinkish skin coloration, a grossly discernible peripheral flush (which persisted up to the time of her discharge).[1] On the following day the patient's temperature began to fall, as did her pulse and respiratory rate; her mental confusion, nausea, vomiting and dyspnea disappeared. This marked detoxification effect was most striking. Forty-eight hours later, pulse, temperature and respirations were normal and the patient was obviously out of danger. She convalesced uneventfully and left the hospital on August 25th, ten days after a single blood irradiation.

---

1 As mentioned elsewhere, a vasodilatation of the vessels of the skin is usually seen very soon after institution of photoluminescent therapy.

Case Studies from Africa

Some of our most revealing and substantive work with photo-oxidation has been done in the disease-racked recesses of Africa. As I mentioned earlier in this book, I've been privileged to be allowed to set up a clinic in Uganda, from which many patients have been treated. The most common disease treated has been AIDS. It is truly an epidemic throughout most of Africa, and my goal has been to offer a treatment that will bring hope to these suffering millions.

* * * * *

After arriving near the very heart of equatorial Africa, we spent two weeks in frustrating opulence at a tourist hotel, waiting to start our great treatment venture. It was worth the wait. Our hosts set us up in a private home with complete security and five bedrooms in which to treat our patients. The house is in a residential area about five miles from the center of town. It was frustrating to waste almost two weeks before getting underway, but the country is desperately short of supplies, and they simply do the best they can. The furniture they brought in was made at a factory the very day it was delivered.

The following case histories are from our AIDS clinic in Uganda. With the positive results that we're experiencing, we're sure the Ugandan government will soon want the world to know of the incredible improvements we're seeing with AIDS, and many other diseases.

Case Histories

N–, John (Bigo), age thirty-four, male CLASS IV
(Our first Patient)
8/14/89:

Occupation: writer, Temp: 37.8°C., Pulse: 100

Weight: about 100 lbs., Height: 6'3"

Some spots in vision. Anorexia, sight of food causes nausea.

Pain at left lower abdominal quadrant (presenting symptom).

Bowels: diarrhea; Urine: o.k.

Cough; but not short of breath.

Lived in Paris: 1982-1986

First Symptoms: Fever, January 1987, and anemia. Was well in two weeks.

Again sick in December 1987; chills for two weeks, then well again.

In July 1988, chills again. In August, violent fever, vomiting for four days, also diarrhea.

Diagnosis of AIDS made August 1988; malaria and typhoid diagnosed, also.

Continued weight loss.

January 1989 – diabetes diagnosed – was in acidosis. Was put on oral diabetic medication. Started gaining weight and felt well after the diabetes stabilized. Family history of diabetes; elder brother is diabetic.

Felt well until April 1989, when abdominal pain returned. Took an herb and got better. Took another herb which brought sugar to normal. Even ELISA returned negative, but Western Blot remained positive.

Early 1989 – syphilis diagnosed. Treated with daily IM penicillin for two weeks – inadequate; treatment repeated, then o.k.

Got sick again in July (early). Now complains only of the abdominal pain, fever, and nausea with vomiting, diarrhea.

Treatment

8/14/89:

8 p.m. H_2O_2 IV

10 p.m. Photoluminescence

No nausea after treatment, no abdominal pain, pulse 100, depressed.

8/15/89:

5 a.m. Photoluminescence

11 a.m. Photoluminescence

Nausea and vomiting returned. No abdominal pain.
4 p.m. Photoluminescence
5 p.m. Temp: 101°F., (37.8°C.), Pulse: 104
No liver tenderness, kept fish dinner down.

8/16/89:
9 a.m. H_2O_2 IV
10 a.m. Photoluminescence
No nausea.
2 p.m. IV vitamins, Mg-1gm., K-20 meq.
Appetite improving. Asking for food. Cheerful. Vomited once.
3 p.m. Photoluminescence
10 p.m. Photoluminescence

8/17/89:
Pulse: 112, Temp: 37.7°C.
8 a.m. Photoluminescence
9 a.m. IV vitamins
10 a.m. H_2O_2 IV
Power out
4 p.m. Photoluminescence
10 p.m. Photoluminescence
Slight diarrhea.

8/18/89:
Retained breakfast
10 a.m. Photoluminescence
11 a.m. IV vitamins.
Hot compress to sore arm.
Now optimistic, "I'm going to get well."
Temp: 37.6°C., Pulse: 104.
Room thoroughly cleaned, given bath, bedding washed.
3 p.m. Photoluminescence
10 p.m. Photoluminescence
Appetite good.

8/19/89:
Ate good breakfast.
9:30 a.m. H_2O_2 IV
11 a.m. Photoluminescence
12 noon Nausea

3 p.m. Photoluminescence
Ate full dinner and retained it.
11 p.m. Photoluminescence

8/20/89:
Severe diarrhea.
Starting oral H_2O_2 10 drops four times a day.
7 a.m. IV Vitamins/minerals
7 a.m. Photoluminescence
3 p.m. Photoluminescence
10 p.m. Photoluminescence

* * * * *

Our author-patient, Bigo N–, after five days of treatment, became dramatically more optimistic and cheerful and said, "I know I'm going to get well."

The next day, as often happens in clinical medicine, our hopes were dashed, as his diarrhea became much more severe. We felt this was due to a yeast infection in his intestinal tract, which is extremely common with AIDS patients in the tropics. I felt that something aggressive had to be done, or we were going to lose our patient from intestinal candidiasis. I made the decision to add oral hydrogen peroxide, three percent, to his regimen, and so began giving him ten drops in a small amount of water as often as he could tolerate it. He received a dose of peroxide orally on the average of every two hours. "We are waiting anxiously for the result," (I recorded in my diary) "and in the meantime, we are giving him intravenous fluids with minerals and vitamins to compensate for his intestinal fluid loss." Two days later, 8/22/89: the diarrhea had completely stopped.

8/21/89:
Ate three eggs for breakfast.
9 a.m. Photoluminescence
3:30 p.m. Photoluminescence
4:30 p.m. H_2O_2, IV and by mouth
Severe diarrhea
9 p.m. IV Vitamins in 250 ml of fluid
10:30 p.m. Photoluminescence
Severe vomiting (caused by oral H_2O_2)

8/22/89:
No diarrhea this a.m.; vein sclerosed, started new IV
8:30 a.m. Photoluminescence
3:00 p.m. Photoluminescence
10:00 p.m. Photoluminescence
500 ml D5W with vitamins
Vomiting continues.

8/23/89:
No diarrhea – H_2O_2 p.o., 8 drops three times a day.
9:00 a.m. Photoluminescence
10:00 a.m. H_2O_2 IV
Retained breakfast
500 ml D5W with vitamins
2:00 p.m. Photoluminescence
Temp: 38.0°C., Pulse: 104
10:00 p.m. Photoluminescence

8/24/89:
Ambulating without weakness around room. Taking balcony visits today.
9:00 a.m. Photoluminescence
Ate breakfast.
H_2O_2 by mouth
IV vitamins = Magnesium, one gm; 'C', 5 gm; B6, 100 mg; Folate, 2 mg.

Follow-up Report From Dr. John B–:

You must be wondering why I didn't start with the report on Bigo. HE IS DEAD. He remained as you left him for quite some time. His main problem was vomiting before eating; fever had gone, also the diarrhea. Somehow, Dr. A- and I decided to give him some "appetizer" (Cyproheptadine tabs, two of them). He became drowsy for the next two days! Couldn't eat. Third day recovered. His sister was about to return from the states so he decided to go back home. (His sister is a nurse.) Taken home on 9/6/89. On 9/18/89, I was called to see Bigo at his house. He was in a critical condition. It was reported to me by his sister (a nurse) that he had had a pneumonia attack, which they were treating with ampicillin injection, 500 mgs every six hours. So I started him on IV H_2O_2, 2.4 cc in one litre D5W, to

run for 12 hours. Repeat dose after four days. He was supposed
to come for photoluminescence as soon as he felt better. DIED
ON 9/22/89.

Bigo's case emphasizes the importance of not stopping therapy too soon. He died 16 days after stopping treatment.[1]

* * * * *

W–, Sam, age 24, male CLASS IV

8/24/89
Occupation: Veterinary assistant
Chief Complaint: Weakness in joints, blurred vision. Fever
intermittently for one month. Disease started in February 1989
with diarrhea, sometimes bloody. Two months later, developed
high-grade fever. Diarrhea severe: as often as twelve times a day.

History of sores and inflammation in mouth, anorexia; no
vomiting. Had typhoid and HIV skin rash, now clearing. Some
dysuria.

Lab:
Hemoglobin - 10 gm on 5/18/89
ESR - 110 WBC - 5300, left shift, toxic granulation
Weidal - Negative
RBC -normocytic, normochromic
  Social: Unmarried, never out of country. Source of AIDS
unknown. Not pursued because of presence of family.

Physical History:
Grossly wasted. Losing hair. No fever, enlarged glands,
thrush [fungus] or skin rash. Resp: Neg chest clear. Abdomen: No
enlarged organs or tenderness. No neurological complaints.
Weight Unknown – scale not available.

Plan:
IV H_2O_2 every other day
Oral H_2O_2 drops - ten (3 percent), three times per day
Photoluminescence three times per day

---

1 Also, as reported in chapter 8, antibiotics block the therapeutic effect of
   photoluminescence.

Follow-up Report from Dr. John B-:

Complained of fever on and off for one month. Now gone. Weakness in joints – now ambulatory. Had lost a lot of weight: 9/5/89 weight was 85.8 lbs., 9/29/89 weight was 88.0 lbs. Had no appetite but now very good. Had diarrhea – now no diarrhea. Painful urination – now no pain. Blurring of vision – vision now clear. Patient wakes up at 8 a.m., takes a bath all by himself. Goes out in the morning sun. He sits up until 2 p.m. in the afternoon. Sometimes he comes downstairs for dinner with us! Completely ambulatory. I have taken photos. A remarkable recovery.

* * * * *

K–, Swaibu, age 48, male CLASS III

8/23/89:

Occupation: Salesman

Started coughing four months ago, productive of white sputum – not foul-smelling. Some chest pain, evening fever; no night sweats. Began to lose his appetite and has had no solid food for two months. Two months ago developed diarrhea, profuse, watery. No blood noted. Six days ago developed skin rash, which itched, and oral sores. Has had nystatin and ketoconazole treatment. Polyuria times six per night.

Social History: Has two wives and 18 children. The first-born is married, last-born is breast-feeding. He has been to Dubai several times and also to Kenya on business.

Physical History:

Pulse: 80

Afebrile, moderate wasting

No lymphadenopathy

Oral candidiasis present

Rash on arms and legs

Chest – clear (x-ray negative)

Heart – NSR

Plan:

HIV Test

Oral H_2O_2, ten drops, three times a day

IV H_2O_2, 3 x per week.

Photoluminescence twice daily; first treatment at 11 a.m. today

Follow-up Report from Dr. John B–:

He has a persistent cough (TB?). Diarrhea has stopped. Abdominal pain has stopped. He's now happier. And has become more involved with his business. As a result, his treatment has become irregular.

Weight: on 8/31/89 was 134 lbs.; On 9/9/89 was 136.5 lbs. Appetite is good.

Patient still has itchy skin rash.

Treatment of Photoluminescence times 21 days, IV H_2O_2, oral H_2O_2, 8 drops four times a day. As I write, I haven't seen him for one week.

Swaibu, we later discovered, decided to go to a witch doctor who advised him to eat dirt in order to keep his disease away. Our treatment, she said, would make him worse. He did not question her as to why he was vastly improved if the treatment was making him worse and so began to eat large quantities of dirt. He soon died from an AIDS-related infection.

* * * * *

B–, Alex, age 27, male, CLASS III

8/24/89:

Occupation: Soldier

Fever intermittently for six months; diarrhea same. Some "pins and needles" sensation, six months.

First Symptoms: weakness, generalized skin rash, diarrhea, vomiting and abdominal pain. Some occasional anorexia.

Sore on penis for seven months (exam – chancre on shaft). Mild, non-productive cough. No chest pain.

Some fever intermittently in evenings.

Treatments for illness have been unsuccessful.

Social: Wife died of AIDS in January of 1989 after a long illness with diarrhea, "slim," and fever. One of their children, age fourteen months, died similarly (presumably with AIDS) four months previously.

Physical History:

Slight wasting – only slight weight loss (8 lbs.)

Afebrile. Pulse – 72

Large, bilateral inguinal lymphadenopathy

Healed M-P rash

ENT – no thrush
Chest/Heart – Negative
Central Nervous System – WNL
Plan: Photoluminescence treatment for three days per week or more;
Oral H_2O_2, drops 8-10, 3 times per day
Monitor weight (present weight unknown)

Treatment

8/24/89:
9 a.m. Photoluminescence
2 p.m. Photoluminescence
6 p.m. Photoluminescence
H_2O_2 by mouth, 3 times per day
Profuse watery diarrhea cleared after two doses of oral H_2O_2/Photoluminescence combination.

Follow-up Report From Dr. John B-:

Fever on and off since June 1st – completely gone.
Diarrhea on and off – gone after a week of treatment.
Sore on the penis for seven months – now dried up and healed! Patient is very happy about it. Within four days the chancre started drying. Appetite improved. Weight on 9/1/89 was 112 lbs.; on 9/7/89 it was 119 lbs. Patient discharged in good condition. Still had the itchy skin rash. No other complaint. Reported back on 9/15/89 for blood checkup.

* * * * *

K–, Francis, age 26, male, CLASS VI

Indigent historian
Was well until seven months ago, when he developed diarrhea, vomiting, and oral sores.
He also developed a fever of high grade associated with rigors. Later, he developed diarrhea and vomiting with general weakness and oral sores. (Sister reports that he developed the skin rash before the onset of all these symptoms.) He was admitted and treated for typhoid. He improved. Given Nystatin ointment for oral sores, chloramphenicol, and Septra. Discharged and stayed home for six months.
In April this year, he developed soreness in the throat, continuing up to now. Has a cough productive of pus-like sputum

with a foul smell. Has associated chest pain. History of diarrhea, but now has stopped. Hasn't eaten any solid food since June (two months).

Passes a pus-like discharge per urethra and has penile sores.

Treatment

Nystatin, Davtrin, Nimorial – no improvement.

He is very confused. Not married. Soldier. Has three children, each with different mother. Eldest six years; youngest four years.

10/4/89: Sed rate 55; Hemoglobin-11.3

1/23/89: Scant malaria parasites

7/25/89: Sputum for TB, none seen
WBC: 3600

6/7/89 - 6/14/89:
Admitted, diagnosis of Pharyngitis, chest x-ray normal.

Admitted to our service on 8/16/89: 2:30 p.m.
Temp 100.4°F., Pulse 120, Respiration 52, shallow
Emaciated, febrile, hot skin. No adenopathy. No skin lesions.
Chest – shallow resp., hyperresonant to percussion; constant, weak cough.
Heart – tachycardia
Abdomen – no organomegaly or tenderness.
2:45 p.m. H_2O_2 IV
3:15 p.m. Photoluminescence
4:30 p.m. Photoluminescence
5:00 p.m. 10 million units aqueous penicillin IV
5:30 p.m. Photoluminescence
6:30 p.m. Photoluminescence
10:00 p.m. Photoluminescence
Patient appears terminal
1:00 a.m. Photoluminescence

8/17/89:
5:00 a.m. H_2O_2 IV

5:30 a.m. Photoluminescence
6:00 a.m. Temp 101°F, Respiration 50
7:10 Patient died.

* * * * *

O–, Anne, age 36, female, CLASS I

8/22/89:
Occupation: banking
Girlfriend of patient, Bigo N–, is asymptomatic.

Plan: Photoluminescence twice a day while living here; then as often as will come in for treatment. Minimum of three treatments per week.
Will give in the muscle because of small veins.

Treatment

8/23/89:
Catheter in vein
Photoluminescence twice

8/24/89:
Photoluminescence twice
Patient did not continue treatment.

* * * * *

Today, three years after the above cases were treated, dramatic changes have occurred in the African clinic. Word of the remarkable improvements seen in terminally ill patients with "slim" has travelled across the country and the clinic is inundated with patients. Record-keeping and laboratory monitoring is practically non-existent but patients look only for results and don't care a banana about T cells and P-24 counts. You don't have to be a microbiologist to see that the treatment is working. When Jobwa comes home with a 30-pound weight gain and his skin rot is gone, the family has seen enough. The brothers, sisters, aunts and uncles are soon lined up at the clinic. A 24-hour wait doesn't matter; they'll pitch a tent in the front yard.

We have proven in Africa, beyond a doubt, that photo-oxidation is the best treatment currently available for the treatment of AIDS. Unfortunately, organized medicine will not accept our findings because of the lack of documentation. This is unfortunate but understandable. What is needed is a million dollars to do a controlled study on Ugandan soldiers with AIDS. AIDS has devastated the Ugandan army and it is a group that can be managed. Lack of proper follow-up is a major problem with African civilians. Excellent laboratory facilities are now available in equatorial Africa and, dropping all semblance of modesty, only lack of money prevents us from demonstrating that photo-oxidation is the best treatment currently available for AIDS and all infectious diseases.

Bibliography

Barnothy, J.: *Biological Effects of Magnetic Fields*, Vol. I., Plenum Press, N.Y., 1964.

Barrett, H. A. *Medical Clinics of North America*, 24: 723, 1940.

Barrett, H. A. "The Irradiation of Auto-Transfused Blood by Ultraviolet Spectral Energy. Results of Therapy in 110 Cases," *Medical Clinics of North America*, 24: 723. 1940.

Barrett, H. A. "The Syndrome of the Posterior Inferior Cerebellar Artery. A. Cinelli." *Archives of Otolaryngology* (in press at time of article, so date unknown).

Barrett, H. A. "Five Years' Experience with Hemo-irradiations." *American Journal of Surgery*, 61:42-53, 1943.

Bayliss, C. E. & W. M. Waites. "The Combined Effect of Hydrogen Peroxide and Ultraviolet Irradiation on Bacterial Spores," *Journal of Applied Bacteriology*, 47, 263-269, Jan. 2, 1979.

Bayliss, C. E. & W. M. Waites. "The Synergistic Killing of Spores of Bacillus Subtilis by Hydrogen Peroxide and Ultraviolet Light Irradiation." *FEMS Microbiology Letters* (in press), 1979.

Bayne-Jones, S. and J. S. Van der Lingen. "The Bactericidal Action of Ultraviolet Light." *Bulletin of the Johns Hopkins Hospital*, 34: 11-16, 1923.

Bellnier, D., K. Ho, R. K. Pandey, J. Missert, T. J. Dougherty: "Distribution and Elucidation of the Tumor-localizing Component of Hematoporphyrin Derivative in Mice." *Photochemistry and Photobiology*, 50: 221-8, 1989.

Beral, V., S. Evans, H. Shore, and G. Milton. "Malignant Melanoma and Exposure to Fluorescent Light at Work." *Lancet*, ii: 290-292, 1982.

Blum, H. F.: **Photodynamic Action and Diseases Caused by Light**. New York: Hafner Publishing Company, 1964.

Bragg, W.: **The Universe of Light**. New York: McMillan, 1933.

Brainard, G. C., M. K. Vaughan, R. J. Reiter. "Effect of Light Irradiance and Wavelength on the Syrian Hamster Reproductive System." *Endocrinology*, 119: 648-654, 1986.

Brainard, G. C., P. L. Pololin, S. W. Leivy, M. D. Rollag, C. Cole and F. M. Barker. "Near-ultraviolet Radiation (UVA) Suppresses Pineal Melatonin." *Endocrinology*, 119, 1986. (In press).

Brainard, G. C., P. L. Podolin, M. D. Rollag, B. E. Northrup, C. Cole and F. M. Barker. "The Influence of Light Irradiance and Wavelength on Pineal Physiology of Mammals." **Pineal-Retinal Relationships**, D. C. Klein and P. O'Brien (editors). New York: Academic Press, 1986. (In press).

Cai, W. M., Y. Yang, N. W. Zhang, et al. Photodyanamic Therapy in the Management of Cancer: An Analysis of 114 Cases, in Kessel D–(ed): **Methods in Porphyrin Photosensitization**, New York, Plenum Press, 1985, pp 13-19.

Carpenter, R. J., N. B. Neel, R. J. Ryan, et al. "Tumor Fluorescence with Hemato-porphyrin Derivative." *Annals of Otolaryngology*, 1977; 86: 661-666.

Chang, T.: Viral Photoinactivation and Oncogenesis. *Archives of Dermatology*, 112: 1176, 1976.

Clark, J. H., Physiological Activity of Light. *Physics Review*, 2: 227, 1933.

Clark, J. H., C. McD. Hill, M. Hanch, J. Chapman and D. D. Donahue. "Ultraviolet Radiation and Resistance to Infection," *Cong. Internat. Lum. Copenhagen*, 458-463, 1932.

Coblentz, W. W. and H. R. Fulton. "A Radiometric Investigation of the Germicidal Action of Ultraviolet Radiation." *Scientific Papers, Bureau of Standards*, 19: 641-680, 1924.

Davidson, Wm. M. "Ultraviolet Irradiation Relative to Anoxia and Bends Susceptibility." *U.S. Naval Medical Bulletin*, 43: 37-38, 1944.

Doiron, Dr. Gomer C. J., S. W. Fountain, et al. Photophysics and Dosimetry of Photoradiation Therapy. Andreoni A, Cubbedu R–(eds): **Porphyrins in Tumor Phototherapy**. New York, Plenum Press, 1984, pp 281-291.

Dougherty, T. J., J. H. Kaufman, D. Boyle, et al. "Photoradiation in the Treatment of Recurrent Breast Carcinoma." *Journal of the National Cancer Institute*, 1979; 62: 231-237.

Dougherty, T. J., J. E. Kaufman, A. Goldfarb, K. R. Weishaupt, D. G. Boyle, A. Mittelman. "Photoradiation Therapy for the Treatment of Malignant Tumors." *Cancer Research*, 39: 2628-35, 1978.

Dougherty, T. J., G. Lawrence, J. Kaufman, D. G. Boyle, K. R. Weishaupt, A. Goldfarb. "Photoradiation in the Treatment of Recurrent Breast Carcinoma." *Journal of the National Cancer Institute*, 622(2): 231-7, 1979.

Dougherty, T. J. "Photodynamic Therapy (PDT) of Malignant Tumors." CRC. *Critical Review of Oncology and Hematology*, 1984; 2: 83-116.

Downes, A. and T. P. Blount. "Researches on the Effect of Light upon Bacteria and Other Organisms." *Proceedings of the Royal Society of Medicine*, 26, 488-500, 1877.

Duggar, B. M. **Biological Effects of Radiation**. Pp. 323-335. New York, 1936, McGraw-Hill Book Co., Inc.

Edell, E. S., and D. A. Cortese. "Bronchogenic Phototherapy with Hematoporphyrin Derivative for Treatment of Localized Bronchogenic Carcinoma: A 5-year experience," Mayo Clinic Proceedings, 1987; 62: 8-14.

"The Effect of Carbon Arc Radiation on Blood Pressure and Cardiac Output." *The American Journal of Physiology*, 114: 594, 1935.

Ellis, C., and A. A. Wells. **The Chemical Action of Ultraviolet Rays**, New York: The Chemical Catalogue Co., Inc., 1925, pp. 236-250.

Felber, T. D., E. B. Smith, J. M. Knox, et al. "Photodynamic Inactivation of Herpes Simplex: Report of a Clinical Trial." *Journal of the American Medical Association*, 223: 289-292, 1973.

Fidler, I. J. "Macrophages and Metastases – a Biological Approach to Cancer Therapy." Presidential Address. *Cancer Research*, 45: 4714-4726.

Finsen, N. R. "Influence of Light on Illnesses," Dissertation at the University of Copenhagen Medical School, 1978.

Forbes, I. J., P. A. Cowled, A. S. Leong, et al. "Phototherapy of Human Tumours using Haematoporphyrin Derivative." *Medical Journal of Australia*, 1980; 2: 489-491.

Fowlks, W. L. "The Mechanism of the Photodynamic Effect." Presented at the Brook Lodge Invitational Symposium on the Psoralens, sponsored by the Upjohn Company, Kalamazoo, Michigan, March 27-28, 1958.

Fowlks, W. L. "The Mechanism of the Photodynamic Effect." *Journal of Investigative Dermatology*, 32: 233-247, 1959.

Ghadiali, Colonel Dinshah P. **Spectro-Chrome Metry Encyclopedia**, Malaga, New Jersey: Spectro-Chrome Institute, Volume 2, p. 431.

Gollnick, K. "Chemical Aspects of Photodynamic Action in the Presence of Molecular Oxygen." In Nygaard OF (ed): **Radiation Research: Biomedical, Chemical, and Physical Perspectives**. New York: Academic Press, 1975, pp 590-611.

Ha X. W., X. M. Sun, J. G. Xie, et al. "Clinical Use of Hematoporphyrin Derivative in Malignant Tumors." *Chinese Medical Journal*, 1983; 96: 754-758.

Hancock, V. K. "The Treatment of Blood Stream Infections with Hemo-Irradiation. Case Reports." *American Journal of Surgery*, 58: 336. 1942.

Hancock, V. K. and E. K. Knott. "Irradiated Blood Transfusion in Treatment of Infections," *Northwest Medicine*, 33: 200, 1934.

Hancock, V. K. and E. K. Knott. "Irradiated Blood Transfusion in the Treatment of Infections." *Northwest Medicine*, 33: 200, 1934.

Hancock, V. K. "Treatment of Blood Stream Infections with Hemo-Irradiation." *American Journal of Surgery*, 58: 336, 1942.

Harris, D. T. "The Action of Light on Blood," communicated to the Biochemical Society Meeting on Dec. 7, 1925, at South Kensington, England.

Harris, D. T. *Biochemistry Journal*, 20: 280, 271, 1926.

"Heliotherapy in the High Alps," *Lancet*, March 19, 1921, 583.

Henderson, B. W. and D. A. Bellnier. "Tissue Localization of Photosensitizers: How Does It Relate to the Mechanism of Photodynamic Tissue Destruction?" CIBA Foundation Symposium on Photosensitizing Compounds: Their Chemistry, Biology and Clinical Use, London, March 1989.

Inamoto, H., et al: Dialyzing Room Disinfection With Ultraviolet Irradiation, *Journal of Dialysis*, 3(263), 191-205 (1979).

Jesionek. Lichttherapie nach Professor v. Tappeiner, *Münch. med Woch.*, 1904, li, 824,965, 1613.

Jodlbauer und v. Tappeiner. "Ueber die Beziehung der Wirkung der photodynamischen Stoffe zuihrer Konzentration," *Münch. med Woch.*, 1905, lii, 2266.

Johnston, R. G. "Monocytes and Macrophages." *New England Journal of Medicine*, 1988; 318: 747-752.

Knott, E. K. and V. K. Hancock. *Northwest Medicine*, 33: 200, 1934.

Knott, E. K. "Development of Ultraviolet Blood Irradiation," *American Journal of Surgery*, August 1948.

Krusen, F. H. **Light Therapy**. New York: Hoeber, 1933.

Laurens, Henry. "Physiological Effects of Radiant Energy," New York, Chemical Catalog Company, Inc., 1933.

Laurens, Henry. "Physiological Effects of Ultraviolet Radiation," *Journal of the American Medical Association*, Dec. 24, 1938.

Lipson, R. L., E. J. Baldes, and A. M. Olsen. "The Use of a Derivative of Hematoporphyrin in Tumor Detection." *Journal of the National Cancer Institute*, 26: 1-8, 1961.

Luckiesh, M., and A. J. Pacini. **Light and Health**, Baltimore: The Williams and Wilkins Co., pp. 66-91.

Macht, D. I. "Contributions to Photopharmacology, and the Applications of Plant Physiology to Medical Problems. Studies on Pernicious Anemia." *Science*, 71: 303-304, 1930.

Macht, D. I. "The Influence of Ultraviolet Irradiation of Menotoxin and Pernicious Anemia Toxin." Proceedings of the Society of Experimental Biology & Medicine, 24: 966-968, 1927.

Manyak, M. J., A. Russo, P. O. Smith, and E. Gladstein. "Photodynamic Therapy," *Journal of Clinical Oncology*; 1988; 6: 380-391.

Mayer, E.: **Clinical Applications of Sunlight and Artificial Radiation**. Baltimore: Williams & Wilkins Co., 1926.

Mayer, E. "Present Status of Light Therapy." *Journal of the American Medical Association*, 98: 3, 221-230, (Jan. 16) 1932.

Mayerson, H. S. and H. Laurens. *American Journal of Physiology*, 86: I, 1928; *Journal of Nutrition*, 3, 465, 1931.

McCaughan, James S., Jr. "Esophageal Carcinoma Treated by Photodynamic Therapy," *Postgraduate General Surgery*, Vol. I, No. 17, (University of Texas), September 1989.

McCaughan, J. S., Jr. "Overview of Experiences with Photodynamic Therapy for Malignancy in 192 Patients." *Photochemistry and Photobiology*, 1987; 46: 903-909.

McLean, R. L. "The Effect of Ultraviolet Radiation Upon the Transmission of Epidemic Influenza in Long-Term Hospital Patients." *American Review of Respiratory Diseases*, 83: 36, 1961.

Melnick, J. L., and C. Wallis. "Photodynamic Inactivation of Herpes Viruses," *Perspectives of Virology*, 10: 297-312, 1975.

Miley, George P. "The Ultra-Violet Irradiation of Auto-transfused Human Blood; Studies in Oxygen Absorption Values," Proceedings of The Physiological Society of Philadelphia, session of April 17, 1939.

Miley, George P. *American Journal of Surgery*, 57: 403, 1942.

Miley, George P. *Archives of Physical Therapy*, 23: 536, 1942.

Miley, George P. *Hahnemann Monthly*, 75: 977, 1940.

Miley, George P. "The Ultraviolet Irradiation of Auto-transfused Human Blood. Studies in the Oxygen Absorption Value." *American Journal of Medical Science*, 197: 873, 1939.

Miley, George P. "The Ultraviolet Irradiation of Autotransfused Human Blood. Studies in the Oxygen Absorption Value." *American Journal of Medical Science*, vol. 14, 1938-1939.

Miley, George P. and J. A. Christensen. "Ultraviolet Blood Irradiation Therapy in Acute Virus-like Infections," *Review of Gastroenterology*. 15: 271-277, 1948.

Miley, George P. and E. W. Rebbeck. "The Knott Technic of Ultraviolet Blood Irradiation as a Control of Infection in Peritonitis," *Review of Gastroenterology*, 10:1, 1943.

Miley, George P. "The Ultraviolet Irradiation of Auto-transfused Human Blood; Studies in Oxygen Absorption Values," *American Journal of Medical Science*, 197: 873, 1939.

Miley, George P. "The Control of Acute Thrombophlebitis with Ultraviolet Blood Irradiation Therapy," *American Journal of Surgery*, 60: 354-360, 1943.

Miley, George P. "Efficacy of Ultraviolet Blood Irradiation Therapy in the Control of Staphylococcemias." *American Journal of Surgery*, 64: 313-322, 1944.

Miley, George P. "Recovery from Botulism Coma Following Ultraviolet Blood Irradiation (Knott technic)," *Review of Gastroenterology*, 13: 17, 1946.

Miley, George P. "Ultraviolet Blood Irradiation Therapy in Acute Poliomyelitis." *Archives of Physical Therapy*, 25: 651-656, 1944.

Miley, George P. "Disappearance of Hemolytic Staphylococcus Aureus Septicemia Following Ultraviolet Blood Irradiation Therapy." *American Journal of Surgery*, 62: 241-245, 1943.

Miley, George P. "Ultraviolet Blood Irradiation Therapy (Knott technic) in Non-healing Wounds." *American Journal of Surgery*, 65: 368-372, 1944.

Miley, George P. "The Knott Technic of Ultraviolet Blood Irradiation in Acute Pyogenic Infections." A study of 103 cases with clinical observations on the effects of a new therapeutic agent. *New York State Journal of Medicine*, vol. 42, No. 1: 38-46, January 1, 1942.

Miley, George P. "Ultraviolet Blood Irradiation: Therapy in Acute Pyogenic Infection at Hahnemann Hospital," *Hahnemannian Monthly*, December 1940.

Miley, George P. "Present Status Ultraviolet Blood Irradiation (Knott technic)." *Archives of Physical Therapy*, 25: 368-372, 1944.

Miley, George. "Ultraviolet Blood Irradiation." *Archives of Physical Therapy*, 23: 536, 1942.

Miley, George P. "Ultraviolet Blood Irradiation Therapy (Knott technic) in Acute Pyogenic Infections," *American Journal of Surgery*, 57: 493, 1942.

Miley, George P. and E. W. Rebbeck. "The Knott Technic of Ultraviolet Blood Irradiation as a Control of Infection in Peritonitis." *Review of Gastroenterology*, 10: 1, 1943.

Miley, George P. and Jens A. Christensen. "Ultraviolet Blood Irradiation Therapy: Further Studies in Acute Infections." *American Journal of Surgery*, 73: 486-493, 1947.

Miley, G. P., R. E. Seidel, and J. A. Christensen. "Preliminary Report of Results Observed in 80 Cases of Intractable Bronchial Asthma." *Archives of Physical Therapy*, 24: 533, 1943.

Miley, G. P., R. E. Seidel, and J. A. Christensen. "Ultraviolet Blood Irradiation in Apparently Intractable Bronchial Asthma." *Archives of Physical Medicine*, 27: 24, 1946.

Noguchi, H. "The Photodynamic Action of Eosin and Erythrosin upon Snake Venom." *Journal of Experimental Medicine*, 6: 252-267, 1906.

Non-linear Electrodynamics in Biological Systems, Plenum Press (1984). eds. W. Ross Adey and A.F. Lawrence. For a discussion of the potential importance of these ideas and the degree to which they have been ignored by the scientific community, see page 595.

Olney, R. C. "Ultraviolet Blood Irradiation in Pelvic Cellulitis." *American Journal of Surgery*, 74: 440-443, 1947.

Olney, R. C. "Ultraviolet Blood Irradiation in Biliary Disease." *American Journal of Surgery*, 72, #2: 235-237, August 1946.

Ott, Nash John. "Color and Light: Their Effects on Plants, Animals, and People," *The International Journal of Biosocial Research*, vol. 7, 1985.

Ott, Nash John: "Color and Light: Their Effects on Plants, Animals and People, part 2," *The International Journal of Biosocial Research*, vol. 8, 1986

Ott, Nash John: **Ultraviolet Light and Your Health**, Devin-Adair Publishers, 6 North Water Street, Greenwich, CT 06830.

Pasternak, B. S., N. Dubin, and M. Moseson. "Malignant Melanoma and Exposure to Fluorescent Light at Work." *Lancet*, i: 704, 1983.

Popp, F. A. G. Becker, H.L. König, W. Peschka (Eds.): **Electromagnetic Bio-Information**, Urban & Schwarzenberg, München-Baltimore, 1979.

Pottier, R., T. G. Truscott. "The Photochemistry of Hematoporphyrin and Related Systems." *International Journal of Radiation and Biology*, 50: 421-452,1986.

Presman, A. S.: **Electromagnetic Fields and Life**, N.Y.: Plenum Press, 1964.

Presman, A. S.: **Electromagnetic Fields and Life**, translated from Russian, Plenum Press, 1970.

Raab. – Ueber die Wirkung fluorescierender Stoffe aufb Infusorien. *Zeitschr. f. Biol.*, 1900, xxxix, 525.

Rahn, Otto. "Invisible Radiations of Organisms," Band IX, protoplasma-monographien, herausgegeben von R. Chambers, et al, verlag von Gebruder Vomgraegen, 1936.

Rahn, Otto: "The Physio-chemical Basis of Biological Radiations," Cold Springs Harbor Symposia on Quantitative Biology, Vol. II, 1934, pp. 226-232.

Rebbeck, E. W. *American Journal of Surgery*, 54: 691, 1941.

Rebbeck, E. W. *American Journal of Surgery*, 55: 476, 1942.

Rebbeck, E. W. and Miley, George P. *Review of Gastroenterology*, Jan.-Feb. 1943.

Rebbeck, E. W. "Ultraviolet Irradiation of Auto-Transfused Blood in the Treatment of Acute Peritonitis, General," *The Hahnemannian Monthly*, April 1941.

Rebbeck, E. W. "Ultraviolet Irradiation of Auto-Transfused Blood in the Treatment of Puerperal Sepsis," *American Journal of Surgery*, 54: 691, 1941.

Rebbeck, E. W. "Ultraviolet Irradiation of Auto-Transfused Blood in the Treatment of Postabortional Sepsis," *American Journal of Surgery*, 55: 476-486, 1942.

Rebbeck, E. W. and R. A. Walter. "Double Septicemia Following Prostatectomy Treated by the Knott Technic of Ultraviolet Blood Irradiation," *American Journal of Surgery*, 57: 536, 1942.

Rebbeck, E. W. "Preoperative Hemo-irradiations." *American Journal of Surgery*, 61: 259-265, 1943.

Rebbeck, E. W. "Ultraviolet Irradiation of Blood in the Treatment of Escherichia coli Septicemia." *Archives of Physical Therapy*, 24: 158-167, 1943.

Rebbeck, E. W. and H. T. Lewis, Jr. (To be published).

Rigel, D. S., R. J. Friedman, M. Levenstein, and D. I. Greenwald. "Malignant Melanoma and Exposure to Fluorescent Lighting at Work," *Lancet*, i:704, 1983.

Riley, Richard L.: "Ultraviolet Air Disinfection for Protection Against Influenza," *The Johns Hopkins Medical Journal*, 140:25-27(1977).

Riley, R. L. and J. E. Kaufman. "Air Disinfection in Corridors by Upper Air Irradiation with Ultraviolet." *Archives of Environmental Health*, 22: 551, 1971.

Riley, R. L., M. Knight, and G. Middlebrook: "Ultraviolet Susceptibility of BCG and Virulent Tubercle Bacilli." *American Review of Respiratory Diseases*, 113: 413, 1976.

Riley, R. L. and S. Permutt. "Room Air Disinfection by Ultraviolet Irradiation of Upper Air." *Archives of Environmental Health*, 22: 208, 1971.

Riley R. L., S. Permutt, and J. E. Kaufman. "Convection, air mixing and ultraviolet air disinfection in rooms," *Archives of Environmental Health*, 22: 200, 1971.

Riley R. L., S. Permutt, and J. E. Kaufman. "Room Air Disinfection by Ultraviolet Irradiation of Upper Air: Further Analysis of Convective Air Exchange." *Archives of Environmental Health*, 23: 35, 1971.

Schneider, Herman and Sperti. "The Quantum Theory in Biology. Institute of Industrial Research," series 4, paper 1, University of Cincinnati (April 2) 1926.

Sieber, F., M. Sieber-Blum. (1986) "Dye-mediated Photosensitization of Neuroblastoma Cells." *Cancer Research*, 46: 2072.

Sieber, F., J. Hilton. Colvin, OH (1984) "Photodynamic Therapy of Drug-resistant Tumors." *Experimental Hematology*, 12: 423 (Abstract).

Sieber, F., S. Rao, S. D. Rowley. Sieber-Blum M (1986). Dye-mediated Photolysis of Human Neuroblastoma Cells: Implications for Autologous Bone Marrow Transplantation." *Blood*, 68:32

Straub. – Ueber den Chemismus der Wirkung belichteter Eosinlösung auf oxydable Substanzen. *Archives for Experimental Patholology*, u. Pharmakol., 1904, li, 393.

Sullivan, M. X., and M. Beroza. The effect of short time ultraviolet irradiation on blood and biochemical compounds. (To be published.)

Sunlight, World Health Publications, Kime, 1980, Box 408, Penryn, CA 95663.

Szent-Györgyi, A. *Science*, 93, 609 (1941).

Szent-Györgyi, A. **Introduction to a Submolecular Biology**, Academic Press (1960).

Szent-Györgyi, A. **Bioelectronics**, Academic Press, 1968.

Szent-Györgyi, A. **The Living State**, Academic Press, (1972).

Szent-Györgyi, A. Electronic Biology and Cancer, Marcel-Dekker (1976). v. Tappeiner – Ueber die Wirkung von Chininderivaten und Phosphin auf niedere Organismen, *Münch. med Woch.*, 1896, xliii, I.

Tappeiner und Jesionek. – Therapeutische Veruche mitfluorescierenden Stoffen. *Münch. med Woch.*, 1903, l, 2042.

Ultraviolet Light and Your Health. *The Swannanoa Health Report*, Issues #2 & 3, Swannanoa Institute, P.O. Box 238, Ivy, VA 22945.

Wallis, C., J. L. Melnick: Photodynamic inactivation of animal viruses: A review. *Photochemistry and Photobiology*, 4: 159-177, 1965.

Ward, H. Marshall. "The Action of Light on Bacteria," Proceedings of the Royal Society of London, v. 54, 1983/1984.

Welch, H. The effect of ultraviolet light on molds, toxins, and filtrates. *Journal of Preventive Medicine*, 4: 295-330, 1930.

Wyckoff, R. W. G. The killing of colon bacilli by ultraviolet light. *Journal of General Physiology*, 15: 351-361, 1932.

Yoshihara, Namiko. "UV Irradiation for Eradicating AIDS Viruses," *Saitama Shimbun*, March 23, 1987.

Zhuralev (ed.), **Bioluminescence of Cells and Nucleic Acids**, English translation, Plenum Press, 1968.

Index